PHASE MEDIA

PHASE MEDIA

Space, Time and the Politics of Smart Objects

JAMES ASH

Bloomsbury Academic
An imprint of Bloomsbury Publishing Inc

B L O O M S B U R Y
NEW YORK • LONDON • OXFORD • NEW DELHI • SYDNEY

Bloomsbury Academic
An imprint of Bloomsbury Publishing Inc

1385 Broadway
New York
NY 10018
USA

50 Bedford Square
London
WC1B 3DP
UK

www.bloomsbury.com

BLOOMSBURY and the Diana logo are trademarks of Bloomsbury Publishing Plc

First published 2018

Library of Congress Cataloging-in-Publication Data
A catalog record for this book is available from the Library of Congress.

ISBN: HB: 978-1-5013-3560-0
ePDF: 978-1-5013-3562-4
ePub: 978-1-5013-3561-7

Cover design by Daniel Benneworth-Gray

Typeset by Deanta Global Publishing Services, Chennai, India
Printed and bound in the United States of America

For Calum

CONTENTS

ACKNOWLEDGEMENTS

The ideas and concepts in this book have partly been honed through conversations with a number of friends and colleagues. In particular I want to thank Lesley Gallacher, Derek McCormack, Sasha Engelmann, Tom Keating, Tom Roberts, Andrew Lapworth, Louise Amoore, Pip Thornton, Agnieszka Leszczynski, Rob Kitchin, Thomas Jellis and Joe Gerlach for various discussions that helped me clarify my position on technology and objects. As my editor, I would like to thank Mary Al-Sayed at Bloomsbury for her quick and professional responses to my queries and for supporting the book at the proposal stage and throughout the production process. In a similar vein, I would like to thank Lesley Gallacher for reading the manuscript and offering helpful advice and comments.

Earlier versions of material that form parts of this book have been presented at a number of academic talks and conferences. Some material from Chapter 1 emerged through a paper as part of a plenary panel for the 7th Annual Doreen Massey Event, held at The Open University in Milton Keynes in 2015. Some material from Chapters 1 and 2 was presented at sessions on geography and post-phenomenology at the Association of American Geographers conference in San Francisco and the Royal Geographical Society Annual Conference in London, both of which took place in 2016. Parts of Chapter 3 were presented at an invited seminar at Kings College in London in 2016. Parts of Chapter 4 were presented at the Royal Geographical Society Annual Conference in Exeter in 2015.

Beyond the specifics of the book itself, I would also like to thank Ken Hillis who generously hosted my visit to the Department of Communication at the

University of North Carolina, Chapel Hill, in 2008. As a geographer, being exposed to a communication studies environment gave me the confidence to begin to think and write across disciplinary boundaries. This book is just one of the outcomes of an interdisciplinary journey that began back when I visited UNC.

1

Phase Media

In the last ten years smart objects have become increasingly ubiquitous. As Wilson (2014) suggests, the use of the term smart to describe forms of digital networked technology and media can be traced back to at least 1995, when AT&T's PhoneWriter Communicator was referred to as a smartphone (Savage 1995). Since then, the term has expanded to describe a whole host of digital devices that are mobile, internet connected and sensor enabled. Examples of smart objects include fridges such as the Samsung Family Hub, whose content can be monitored while shopping via a smartphone app; watches such as the Apple Watch, that measure heart rate; tooth brushes such as the Philips Sonicare that measure and visualize the duration of brushing; phones such as the Google Pixel that track location; television remotes such as the LG Magic that are voice activated; and light bulbs, such as the Philips Hue that respond to ambient light conditions, among many others. Depending on their particular sensor components, smart objects can sense a range of environmental information, such as motion, temperature, location, light, air quality, vibration, pitch, elevation and environmental sound, and relay this information to other devices, such as laptops, phones, centralized servers or control rooms.

As part of an *Internet of Things* (Greengard 2015), these objects are described in different contexts as locative (Tuters and Varnelis 2006; Licoppe 2013), geo (Crampton et al. 2013), spatial (Leszczynski 2014), mobile (Farman 2012), pervasive (McCullough 2004), ubiquitous (Galloway 2004), sensing

(Gabrys 2014), context aware (Schmidt 2014), environmental (Parikka 2015), ambient (Crang and Graham 2007) and so on. The desire to label these diverse technologies is driven by a belief that there is something distinct about smart devices as a sub-category of technology and media more generally. The terms introduced above aim to get at this specificity in different ways. For example, the phrase locative media is used to refer to the fact that so much activity 'gets tracked, tagged and mapped. Cell phones have become location-aware, computer games have moved outside, the web is tagged with geospatial information, and geobrowsers like Google Earth are regarded as an entirely new genre of media' (Thielmann 2010: 2). Services and apps such as Flickr and Foursquare can only work when accessed from smartphones that can communicate their spatial location and these devices have become pervasive and ubiquitous throughout the world (on the history of ubiquitous computing see: Galloway 2004; Kinsley 2011, 2012). Beyond items embedded in the home or carried by individuals, many governments and businesses are now attempting to create smart cities, where information from a variety of sensors such as cameras and motion detectors in buildings and on streets are collected and analysed in order to manage urban spaces.

On one level, the term 'smart' could be easily dismissed as a marketing jargon used to advertise and sell a range of digital goods. Writing for technology magazine *Wired*, Clarke (2015: n.p.) encapsulates the hyperbole surrounding smart objects nicely:

> We are all constantly bombarded by … technology trends but there are two genuine technology tsunamis heading our way right now, namely the Internet of Things and Smart Machines. Where these two forces collide with one another and with us, they will create an explosion of new opportunities in areas as diverse as entertainment, healthcare, disaster management and smart cities. If you thought the mobile revolution has had a huge impact on individuals and businesses, you ain't seen nothing yet!

Smart objects are marketed as providing convenience and efficiency for consumers, policymakers and governments alike, promising the ability to control and track various aspects of people and things more accurately. For citizens the advantages of this tracking are promoted as an opportunity to become more efficient, healthier subjects. For instance, smart thermostats are designed to save money on domestic heating, and fitness trackers are designed to encourage users to get more exercise. For governments, the promise of smart cities is that drawing together data from a city in real time will help alleviate all manner of problems associated with urban environments, such as traffic congestion and poor air quality (Kitchin 2011; Batty 2013; Campbell 2013). Underlying these discourses of ease and convenience is the fact that developing smart technology and media is incredibly lucrative from a business perspective. The smart wearable technology market is estimated to be worth $30 billion by 2020 (Hunn 2015) and the global market for smart city solutions across the sectors of water, transport, energy, waste and assisted living is valued at $400 billion within the same period (BIS 2013).

Despite the utopian and idealistic discourses that are used to sell and advertise smart objects, responses to this technology have tended to centre on a critique of the way they enable and reinforce existing forms of corporate power and control. As Silverman (2016: n.p.) puts it, 'Smart produces a world where we no longer exert control over objects we've brought from corporations, but corporations exert control over us through things we pay for the privilege of using. And when "smart" is crudely applied to the cities we live in … we give in to forces of privatization, algorithmic control and rule by corporate contract.' This concern is reflected by a range of writers who work across media, and cultural, science and technology studies, as well as disciplines such as geography. For example, Dodge and Kitchin (2009: 1361) suggest that the kind of control that smart objects enable can

> vary from subtle, almost voluntarist, conditioning that is little noticed, such as the pre-selection of potentially interesting programmes to watch

> by a television, or `body monitoring' bathroom scales chiding the user for missing their target weight and urging greater efforts of performance … to a more potent form where coded objects refuse to perform because they determine that an action is 'illegal' (for example, copyright enforcement through digital-rights management stopping the computer playing movies not legally owned).

Expanding this argument, writers such as Andrejevic (2007) and Coley and Lockwood (2012) point to networks of smart objects creating a digital enclosure, where action is monitored and controlled at every moment. Coley and Lockwood (2012 : 22) discuss the Google search engine as predicting and thus conditioning what people are looking for, 'permitting only certain moves to be made … [and] … refusing others'. In their words, these 'network structures perpetuate a new, immanent control, through sets of rules and formulae, codes, which pre-format our actions and behaviours' (Coley and Lockwood 2012: 17). In doing so, these forms of control contribute to capitalist processes and 'system[s] of value production that … produce profit only by exploiting and dispossessing human life' (Franklin 2015: xviii). While these critiques are undoubtedly important, they often focus on how smart objects affect human life and behaviour. The objects themselves tend to be treated as tools or mediums through which control or regulation is enabled. For Coley and Lockwood, what matters more is the type of power smart objects, connected via the cloud, enable, rather than the specificities of the smart objects that actually make up cloud computing services.

If we focus on smart objects themselves, however, a stranger picture emerges. Rather than spaces and times organized by smart objects to suit various corporate or government interests, we encounter a range of complex, normally imperceptible overlapping forces and fields that would seem to exceed the control of any one company or individual. Communicating via radio waves, Bluetooth, ultrasonic sensors, light detection and a variety of other signals, smart objects regularly 'dive into the nooks, crannies, gaps and wormholes that exist in an imperceptible and often invisible world that extends beyond human eyes, ears, smells and consciousness' (Greengard 2015: 22). This is an important zone to explore because, as Harman (2016: 6) argues, 'if we forget that objects interact among themselves even when humans are not present' then we ignore much of what these objects are doing in the background. In the case of smart objects, this background activity can shape how people make sense of the world, by 'affect[ing] how people arrive, depart and inhabit … [places and] … how they relate to others' (Mackenzie 2010: 5).

Focusing on these often imperceptible background forces, this book argues that what smart objects do should primarily be understood in terms of the qualities they disclose outside of the realms of human sense. Focusing on the mechanisms and operations of disclosure, I argue smart objects generate phases. For now, phases can be defined as the spatio-temporalities disclosed by smart objects. Despite (or because of) their imperceptibility, phases modulate the intelligibility of space and time for the humans and non-humans who find themselves alongside them. As we shall see, phases modulate space and time by disclosing qualities of objects that create relations of near and far and then and now, according to a variety of logics. In relation to phase spaces, these logics include diffusion, partition and envelopment. In relation to phase times, these logics include gradation, dispersion and dilation. Although these logics are regularly utilized for commercial purposes in an attempt to modulate space and time to structure consumer demand and regulate users' behaviour, I will argue that smart objects cannot be reduced to these logics, or the instrumental goals of their creators.

This is because phases, and the logics that underlie these phases, shape the potential for human action and the kinds of political activity that emerge through them in ways that cannot be entirely controlled or anticipated by the manufacturers and designers of smart objects. From this position, arguing that smart objects 'conform … to a conservative political agenda that represents citizens as automated/autonomous subjects, ideally engaging in self-responsibilized practices of dataveillance and life optimization and emitting valuable "data exhausts" for repurposing by other actors and agencies' (Lupton 2016: 118) misses how smart objects regularly create phases between things and people that are indifferent to and confound the logics of the venture capitalists that fund and run the companies that create these objects. While phases attempt to structure consumer demand and behaviour, they can just as easily create intelligibilities that challenge or undo the very phases that created these intelligibilities.

Emphasizing this point is important. While there are many excellent critiques of smart objects available (which are discussed in detail in Chapter 6), this book does not want to offer just another denunciation of smart objects as a tool of control and oppression. Rather, I want to focus on what smart objects do and the space-times they disclose, which can never be fully determined by their creators. Focusing on the contingency of smart objects provides tools for responding to phases without becoming politically despondent or hopeless. This is a political position that understands politics as shaped by the particular objects that make up 'circumstances of the time', but 'it also understands those circumstances as new material that can be forged into fresh movements and alliances and into new inventions of what the political itself consists of' (Amin and Thrift 2013: 8; Gerlach and Jellis 2015). The smart objects I analyse, and the logics that underlie them, are thus sites of new forms of political contestation as much as they are emblematic expressions of corporate or capitalist power.

Formally put, as I will unpack in Chapter 6, the book recognizes what could be termed, following Jean-Luc Nancy and Aurélien Barrau (2015), the constructive as well as structive logic of smart objects. While construction refers to the instrumental goals smart objects are designed to realize, struction refers to the way smart objects exist alongside one another and disclose phases that always exceed their design. The examples discussed across the book – from light bulbs, to voice assistants, to autonomous vehicles – have therefore been explicitly chosen to point to how phases can be generated to create profit and modulate behaviour for commercial ends, but whose same logics can also be utilized to address complex social problems and issues. The concept of phase, therefore, allows us to analyse the constructive and structive logics of smart objects and attend to both the problems and potentials of these objects.

To situate these arguments and claims, the rest of this chapter is split into the following sections. The section 'Networks' suggests that the concept of phase alters how we think about networks beyond notions of flat or vertically organized points of capture and translation connected by lines of transmission. The section 'Smart Objects' argues that in order to understand smart objects as producing phases, we need to develop an object-orientated approach to smart devices as part of a broader category of technology. The section 'Smart Objects, Space and Time' unpacks the relationship between phases and space and time using debates around process to contextualize the phase approach to objects developed in the book. The section 'Exploring Phases' looks ahead to the rest of the book and provides concise summaries of the key arguments in each of the chapters and the examples used to make these arguments.

Networks

Alongside a focused examination of smart objects, the concept of phase is designed to speak to broader literatures around technical networks and the concept of networks more generally. Specifically the notion of phase provides a new imaginary for thinking about what networks are, both within and beyond a purely technical definition, and opens up new understandings of politics that are associated with the term. This is an important task because the concept of the network has become absolutely central to understanding digital objects in general and smart objects in particular. As Galloway and Thacker (2007: 25) suggest, 'For the last decade or more, network discourse has proliferated with a kind of epistemic intensity: peer-to-peer file sharing networks, wireless community networks, terrorist networks, contagion networks of biowarfare agents, political swarming and mass demonstration, economic and finance networks, online roleplaying games, personal area networks, mobile phones, generation Txt and on and on.'

Writing in *Protocol*, Galloway (2004: 3) suggests networks can be defined as 'a structural form without center that resembles a web or meshwork'. From this perspective, networks are nodes or objects that are connected to one another in some form. More specifically, Galloway and Thacker (2007: xv) define computer networks such as the internet as 'a set of technical procedures, for … managing, modulating, and distributing information throughout a flexible yet robust delivery infrastructure'. Here networks are understood as anti-hierarchical, existing on a flat plane, where one object has the potential to connect to any other.

Alongside such flat approaches, writers such as Bratton (2016) have argued for an understanding of technical networks as a stack of vertically layered elements. Bratton (2016: 52) defines the stack in the following words: 'The stack … is a vast software/hardware formation … built of crisscrossed oceans, layered concrete and fibre optics, urban metal and fleshy fingers, abstract identities and the fortified skins of oversubscribed national sovereignty. … Its properties are generic, extensible, and pliable; it provides modular recombinancy but only within the bounded set of its synthetic planes.' Regardless of whether technical networks are understood as either flat or vertically layered, these accounts are organized around the fundamental logic of a net or mesh. This notion of networks as a web or meshwork extends beyond narrowly technical definitions of computer infrastructures. Within philosophy and the social sciences, as Vitale (2014: 16) argues, pretty much anything can be understood as a network, if we consider a network – 'any whole, composed of parts, distinguished from a background, and composed of other parts and wholes, layered into each other at multiple levels of scale'. Here the term 'network' becomes a kind of meta-concept for understanding all aspects of the world. For Vitale (2014: 3), 'If there is one word which brings together the multiform new logics which are so rapidly changing the structure of our world, a word which describes the ways in which everything is fracturing so as to reconnect more intensely, it is the term "network".' Indeed Actor Network Theory (ANT), one of most important social theories to emerge in the last thirty years, also draws upon notions of networks as webs of connection that are more or less durable to understand social reality. For Latour (2005: 132) a network is, among other things, 'a point to point connection … which is physically traceable', 'a connection … [that] … leaves empty most of what is not connected', which 'is not made for free … [and] … requires effort'. While Latour and other ANT scholars such as Law (2002, 2004) and Müller and Schurr (2016) are keen to emphasize the processual, dynamic nature of networks, the notion of the network as a web or meshwork is still present. Rather than a stable thing or substance, networks become 'the trace left behind by some moving agent' (Latour 2005: 132).

Despite their popularity, Galloway and Thacker argue that these meshwork theories of networks, in relation to technical objects in particular, need to be challenged. As they put it, understanding networks as webs 'excludes what makes a network a network (its dynamic quality), but … also require[s] that networks exist in relation to fixed, abstract configurations or patterns … and to specific anthropomorphic actors' (Galloway and Thacker 2007: 32). In other words, they argue that understanding networks as webs is inadequate in the sense that networks can appear too static and rigid and the role that non-human forces and objects play in their operation can be minimized.

To remedy the weaknesses of meshwork approaches, Galloway and Thacker gesture towards an alternative imaginary of networks, based around the elemental. According to Galloway and Thacker (2007: 157) the elemental refers to the dynamics of networks that operate above and below human subjects: 'The elemental is this ambient aspect of network, this environmental aspect – all the things that we as individuated human subjects or groups do not directly control or manipulate.' These environmental aspects could include the non-technical objects and processes that nonetheless shape how networks operate, such as meteorological forces like lightning that interfere with radio communications or aquatic life that damage undersea internet cables (Engelmann 2015; Starosielski 2015; McCormack 2017). For Galloway and Thacker (2007), developing this elemental account 'requires us to elaborate an entire climatology of thought … in terms other than the … overly spatialized dichotomy of nodes and edges' that proliferate when theorizing networks as a set of points and lines.

Developing Galloway and Thacker's argument, I suggest that the concept of phase provides an alternative to a nodes and edges account of networks that proves useful both within and beyond an examination of smart objects. As will become clear across the following chapters, key to the notion of phases is that smart objects cannot be reduced to the effect or outcome of a network. Following Martin Heidegger and Gilbert Simondon, the book argues that smart objects have an intentionality (an ability to actively perturb other objects) and protentionality (an ability to await to be perturbed by other objects and respond to these perturbations). In turn, rather than fixed nodes or points connected by lines or means of transmission, smart objects' perturbations create diffused phases, in which particular qualities of objects are disclosed and shift into and out of appearance, generating space-times that are localized to them.

Smart objects

To theorize phases requires that we move away from a relational or process-based notion of what a smart object is. Understanding objects as processes or sets of relations has become popular across the social sciences and humanities.

Drawing upon figures such as Deleuze (1988a,b), Whitehead (2010) and Barad (2003, 2007) among others, movements such as assemblage theory (Anderson and McFarlane 2011; DeLanda 2016), new materialism (Bennett 2009; Washick et al. 2015) and aspects of speculative realism (Bryant et al. 2011, Shaviro 2014) have emphasized 'the dynamic, temporal and process character of systems and things' (Connolly 2013: 400). From this position, we are told that sets of objects are actually changing forces and intensities of matter. For instance, Amin and Thrift (2017: 2–3) define cities as 'spatial radiations that gather worlds of atoms, atmospheres, symbols, bodies, buildings, plants ... and institutions, each with its own mixes ... [and] ... moorings ... held in relation within specific networks'. From this perspective, as Coole and Frost (2010: 9) put it, 'Materiality is always something more than "mere" matter: an excess, force, vitality, relationality, or difference that renders matter active, self-creative, productive, unpredictable.' In emphasizing the active, amorphous nature of matter, discrete objects can become positioned as the more or less illusory or temporary outcome of a process that only ever appears to be stable for living beings due to the peculiarity and specificity of different biological and phenomenological systems of sense perception.

Process-based accounts of objects often make an associated claim around the role that relationality plays in constructing the appearance of entities as discrete or distinct (Roberts 2012; Lapworth 2015; Williams 2016). New materialist perspectives regularly suggest that relations precede objects and it is relations or processes that create an appearance that objects or categories of thought are discrete or distinct. As Del Lucchese (2009: 181) puts it, from a process perspective, 'Being is not what "is" (and what eventually happens in the form of relations); Being is what "becomes" in and through relationality.' In Fraser et al.'s (2005: 3) words, 'In process thinking, relations and relationality cut through and across all spheres, regardless of the distinctions that are drawn between them (between the cultural, the natural, and the artificial, for example). Indeed, distinctions are themselves an aspect (an effect) of the differentiation of processes.' This understanding is reflected in the work of Parikka, who suggests that media networks can be understood as continua of objects that cannot be categorically distinguished from one another as either natural or artificial. In his words, media networks 'remain irreducible to either their "hard" contexts … [such as] … CO, toxic materials, minerals, and other component parts … or their "soft" bits (signs, meanings, attractions, desires)' (Parikka 2012: 97). From this flat perspective, 'signs are transmitted as signals, through cables, in hardware, in a mesh of various components from heavy metals to PVC coatings. … In short: continua all the way down (and up again); soft to hard, hardware to signs' (Parikka 2012: 97).

Here objects are considered to be ultimately interconnected, with a small change in one object having the potential to generate 'massive but unanticipated effects' (Coole and Frost 2010: 14) in whole range of others. Coole and Frost use the example of the so-called 'butterfly effect' to explain this. The butterfly effect 'refers to the possibility that a slight disturbance of air precipitated by a flapping of diaphanous wings could set off a succession of complex meteorological and atmospheric changes that trigger a hurricane in another hemisphere' (Coole and Frost 2010: 14). From this position, all objects form part of a holistic network in the sense that one interaction among two entities has the potential to transform all the objects in the network as a whole.

The unpredictability of object relations is often explained through an emphasis on the event as a moment of generative encounter. In Fraser et al.'s (2005: 4) words it is 'the event … [that] … makes the difference …. Importantly, motion and change are attributable to differences within an event. Duration is the field for the event: there are as many durations as there are events. The accent on process … corresponds to a privileging of an ontology of becoming over being – to a flux that cannot be understood except in terms of process, or passage.' Drawing upon the work of Urry (2005: 5), Coole and Frost

(2010: 14) go on to argue that this evental model of relationality results in an account whereby

> "there are system effects that are different from their parts… . [The] … components of a system through their interactions spontaneously develop collective properties or patterns … . These are non-linear consequences that are non-reducible to the very many individual components that comprise such activities". Because innumerable interactions between manifold elements that produce patterns of organisation successively transform those elements, it is impossible either to predict outcomes in advance or repeat an event.

Arguments from new materialism and assemblage theory that focus on the relational and evental nature of objects are exciting in the sense that they make banal or supposedly static things active and dynamic again. But, while they can lead to fascinating thought experiments, if we take the processual, interconnected and evental nature of every object seriously a number of issues arise. For the interests of this book, we can identify and focus on two such issues.

First, a focus on relationality can generate what Culp (2016) terms a 'connectivist' form of thinking. If we accept that everything is ultimately connected to everything else in some format or on some level it soon becomes difficult to differentiate between one object and another at all. In turn, faced with the overwhelming complexity of a system or network, connectivist thinking can lead to a kind of critical paralysis, because any action might lead to disastrous effects for some unknown or non-present other. As Morton argues, 'consider how we're now aware of risks on global and micro scales. We can find out exactly how much mercury our bodies contain. We know that popular kinds of plastic leach dioxins', but 'the more risk we know about, the more risk spreads. … Along with a sense of tremendous power … of being able to see everything … goes a sense of perilous vulnerability' (2010: 25).

Second, in arguing that relations precede objects or that objects are constructed through relational events, the very phenomena under study can disappear from view because they can be reduced to a set of processes, relations or events that supposedly produce them. In other words, the process or relation, understood as an event, becomes the focus of analysis, rather than the objects and bodies involved in that event or encounter. As Alaimo (2013: 15) puts it,

> Contemplating the utterly counter-intuitive sense of the world as made up of intra-acting agencies, rather than separate objects, is to me, more dizzying and generative than contemplating objects as distinct, alien beings. Tracing intra-actions and other modes of entanglement between substances and systems enables political critique and the development of ethical and political modes that do not separate the human from the material world.

While a sense of dizziness may be aesthetically pleasing or disruptive for the researcher, it would seem difficult to develop 'ethical and political modes' of response to problems when the very object that is the supposed focus of the investigation cannot be clearly identified.

Recognizing these issues, the book draws upon ideas from speculative realism (Harman 2010b, 2013), post-phenomenology (Lea 2009; Ihde 2010; Ash and Simpson 2016; Ash and Gallacher 2015) and mechanology (Simondon 2017). Developing this work, the book argues that objects are not sums or lists of properties that can be added together to provide a total account of a thing but nor are they processual becomings. Instead objects are defined as units that exist 'prior to their relations or effects' (Harman 2016: 42–3). Here objects are concrete units that can only ever be apprehended by what they do, but are never reducible to what they do. In other words, objects have an internal core that is never completely displayed or exhausted regardless of how they are probed or investigated. But, different from new materialist positions, the generativity of the object does not lie outside of the object as an event, relation or process. Instead the generativity of the object is always internal to that object or its components. To analyse an object you need to stick with the object itself rather than turn to a set of relations or processes that precede that object in order to explain what it is or what it can do. Developing this definition of objects in general, we can state that smart objects are units that cannot be reduced either to how they are manufactured or used or to non-human or vital events, processes or relations that precede them. Instead, smart objects, as part of a broader category of technology can be defined by their intentionality and protentionality.

As will become clear over the rest of the chapters, the account of objects developed in the book does not eschew or deny that smart objects enter into relations with other things, or the fact that relations are important for what smart objects can do. As Schwanen (2015: 677) argues in relation to smartphone software apps,

> It would be patently simplistic to locate all of the forces or effects such artefacts as apps are able to generate in those objects themselves. Anyone used to handling physical mobility-related apps knows that most owe their usefulness and effectiveness to the ability to provide real-time, location-based information, which is only possible if they are hooked up to wider constellations for data exchange and communication.

A phase analysis of smart objects does not reject relational thinking per se, but rather works to rethink the status and place of relationality as an explanatory term to account for what smart objects are and what they do. From a phase perspective the appearance of relationality is the outcome of the way objects perturb one another and disclose qualities, rather than objects being the product of relations. To practice this kind of inversion, in Schwanen's (2015: 677) words, is a challenge that involves understanding how the smart objects that make up a phase 'generate effects in and of themselves on the one hand, and appreciating the relations with other entities in which they are situated on the other'. As I go on to argue in Chapter 2, focusing on the objects themselves is the key to understanding how phases are constructed and the effects they have on both humans and non-humans.

Smart objects, space and time

This book argues that smart objects produce phases, or space-times, around which human and non-human life is organized. The notion that technologies contribute to the production of space and time is not new. Within geography for instance, writers such as Harvey (1992) have long theorized how technologies produce time–space compression and distantiation in order to understand phenomena such as globalization. More recently, work in digital geographies (Ash et al. 2016) has sought to understand how spaces such as cities are mediated as 'code-space' (Kitchin and Dodge 2011) by various digital devices and smart objects and how these code-spaces alter economic (Graham and Marvin 2001; Richardson 2016; Bissell and Del Casino 2017), social (Elwood 2006) and cultural (Elwood and Leszczynski 2011; Rose 2015) practices and exacerbate existing forms of inequality (Zook and Graham 2007; Gilbert 2010; Wilson and Graham 2013).

In parallel to these debates, within media and cultural studies a number of writers have begun to think about the relationship between digital devices and space and place as well (Goggin 2012; Goggin and Hjorth 2014; Pink and Leder MacKley 2013; Bertel 2016; Pink and Fors 2017; Wilmott 2017). Here, digital devices have regularly been understood to link together aspects of physical and digital space to create so-called 'hybrid spaces' (de Souza e Silva and Delacruz 2006; e Silva 2006; Frith 2013), or create new convergent experiences of space (Wilken 2008; Gazzard 2011; Wilmott 2016). At the same time writers such as Humphrey (2010) point to the potential for smart objects to alter how people come together and use existing public spaces such as parks in new ways.

Despite the differences in these approaches, what unites the thinkers above is that their work utilizes a more or less relational framework for understanding space and time. Relational space can be understood as a turn away from thinking space as a fixed container in which action takes place. As Jones (2009: 491) argues, a relational approach to space is a

> paradigmatic departure from the concerns of absolute and relative space, because it dissolves the boundaries between objects and space, and rejects forms of spatial totality. Space does not exist as an entity in and of itself, over and above material objects and their spatiotemporal relations and extensions. In short, objects are space, space is objects, and moreover objects can be understood only in relation to other objects – with all this being a perpetual becoming of heterogeneous networks and events that connect internal spatiotemporal relations.

From a relational perspective, space and time become wedded together. In Murdoch's (2005: 74) words, 'There is no absolute space (just as there is no … absolute time); only specific space-time configurations … conditioned by … relations.' In geographers' work on digital objects this relational influence is clear. As Dodge and Kitchin argue, code-space is fundamentally relational, whereby 'space is not simply a container in which things happen; rather spaces are subtly evolving layers of context and practices that fold together people and things and actively shape social relations' (Kitchin and Dodge 2011: 13). In work on mobile media in media and cultural studies, a relational perspective is echoed by writers like e Silva who suggest that mobile phones 'not only reshape communication relationships but also reshape the space in which this interaction takes place' (e Silva 2006: 262).

While accepting Jones's point that space and time do not exist over and above objects, a phase account of smart objects differs from a relational account of space developed in geography and employed in work on digital technology and smart objects. The key point of difference for a phase approach is that space and time do not emerge solely from the way the external features of these objects encounter one another through a 'perpetual becoming of heterogeneous networks and events that connect internal spatiotemporal relations' (Jones 2009: 491). Space and time are not produced through an event or moment of contact in which the qualities of an object are actualized from elsewhere, but become disclosed as an object exposes qualities that are internal to that object. To understand space and time is to understand how space and time appear as those qualities disclosed by the way an object perturbs or is perturbed by other objects. The qualities these perturbations disclose are not solely the product of the relation between objects, as in process or relational approaches, but exist in the objects themselves.

Such a difference in position may appear subtle at first glance, but in practice results in a very different way in accounting for and analysing smart objects. In particular, through a phase analysis of smart objects, space and time come to be defined by the accessibility of objects rather than quantitative units, distance, geometries or relations. Accessibility is a term that emerges from the work of Heidegger (see Dreyfus 1991 in particular Chapter 7). The phenomenon of accessibility can be illustrated through the simple example of a book on a shelf. Imagine a book on a high shelf 4 metres in front of you. Without a ladder to reach the book, the book is further away from you than a book lying at the other end of the room 10 metres away, because you cannot access it. From this perspective, distance is qualitative and dependent on how objects perturb one another, rather than purely a matter of quantitative geometric measure (Heidegger 1962). Of course, when Heidegger discusses spatiality through the phenomenon of accessibility it is in specific reference to a human being who is moving around a space for some purpose. As Schalow (2000) and Glendinning

(1996) among others argue, Heidegger (1995: 196) denies the capacity for accessibility to non-human things or even animals, outside of some brute form of causal encounter. The account of phases developed across the rest of the chapters rejects this denial. Phases are precisely about how objects access one another through the qualities they disclose outside of modes of human perception or processes of brute causality. Distinct from Heidegger, by accessibility I mean that what constitutes space is when qualities of an object create the appearance of distance or nearness to another object. What constitutes time is when one object perturbs another and discloses qualities that create the appearance of change. Indeed, phases emerge specifically from the overlapping, disjunctive modes of disclosure of qualities as smart objects perturb and are perturbed by other things. As I argue in Chapters 3 and 4, smart objects can be understood as units of shifting qualities that are multidimensional. Smart objects are not simply in relation or out of relation with one another or other objects in binary terms. Instead, different smart objects can overlap one another and create the appearance of multiple space-times that do not necessarily intersect or meet.

Exploring phases

The rest of the book explores these arguments and themes by utilizing the following structure. Chapter 2 develops the notion that smart objects can be understood as different in degree from other technical objects. Smart objects have an intentionality and protentionality that is shaped by the perturbations that their allopoietic and homeostatic structure partly affords. To discuss intentionality and protentionality, I draw upon a number of examples, including the Apple Lightning Cable and the Apple Siri voice-controlled personal assistant. These examples allow us to understand what smart objects can do, how they differ from other technical objects, and why they can't be reduced to their instrumental design and manufacture. In turn this definition provides the groundwork for understanding how smart objects generate phases.

Chapter 3 develops the concept of phase spaces. When smart objects perturb one another they disclose or fail to disclose a series of qualities that generate relations between near and far, which modulates how space appears. The chapter identifies three logics of spatial modulation that it terms diffusion, partition and envelopment. The chapter unpacks these logics in relation to smart objects including the Sun weather app, the Nest Security Cam and the Lumo Run smart running tracker. Through these examples, the chapter argues that phase spaces operate outside of the realms of phenomenal experience but still work to structure and prime spatial activities and practices.

Chapter 4 suggests that smart objects also generate phase times, understood as the processes of perturbation between smart objects, through which relations between then and now become intelligible within phases. Developing the examples of smart items such as the Drift light bulb, the Dyson 360 Eye vacuum cleaner and the Apple Watch HeartWatch app, I show how these items disclose temporal phases that modulate the appearance of time according to logics of gradation, dispersion and dilation.

Chapter 5 reflects on how the concept of phase alters political questions around the nature of smart objects in everyday life. Discussing Harman's introduction of the distinction between up and down politics, alongside traditional categories of left and right politics, the chapter suggests taking smart objects seriously asks us to supplement these distinctions with notions of exo and endo politics. Rather than ontologizing politics, which leads to a situation where every entity is potentially political, an exo and endo politics works to identify the conditions through which political issues appear in a situation. To show how political issues appear in relation to smart objects and the objects that support smart objects, the chapter examines the case of the Wishaw phone mast dispute in the UK. Identifying exo and endo politics as legitimate areas of concern, the chapter suggests that becoming sensitized to the phases of smart objects is the key to enabling new technical politics to develop.

Phases can be generated for all kinds of purposes, some of which are negative for the humans and non-humans who find themselves within them. How can individuals, organizations and collectives respond to the logics of phases that companies and governments attempt to impose? Chapter 6 suggests that the subversion of problematic phases is possible and terms this subversion involution. Involution is a process through which smart objects are altered to disclose qualities that cause a phase to become reorganized or dismantled, opening the possibility for new phases to emerge. Through case studies of the Uber ride-sharing app and the Umbrella Movement in Hong Kong, the chapter shows how such involutions are possible. Unpacking the notion of involution, the chapter differentiates this type of engagement from existing understandings of critique that are usually associated with political intervention.

In the final substantive chapter, the books turns to the potential ethical issues associated with the rise of smart objects. Discussing the example of semi-autonomous vehicles such as the Tesla Model S and autonomous vehicles such as the Google car, the chapter forwards a notion of phase ethics. Through the analytical lens of the accident, the chapter shows how a phase ethics can be usefully applied to reflect on the promises and problematics of smart objects that operate with little human oversight. The chapter suggests that a phase ethics of smart objects is particularly important because of their increasing ability to cause harm and death depending on how they respond to other objects. In turn the chapter offers some practical recommendations around the safety of autonomous vehicles to demonstrate how a phase ethics can be used to intervene in political debates around smart objects.

To conclude, the book turns to the broader concept of networks and reflects on how an understanding of phases alters how we think about and conceptualize networks. Rather than points and lines, we can consider networks as ambient phases that cross and intertwine with all manner of non-technical objects. In turn, the conclusion suggests phases may become denser and more tightly folded into human and non-human intelligibility as the number of smart objects increase across the global North and South.

2

Objects

Smart objects cannot easily be defined using long-standing, pre-existing definitions of technology. Writing in 2006, Kroes and Meijers (2006: 1) usefully summarize one such definition: 'Technical artefacts such as typewriters, hammers, copying machines or computers are different from social artefacts such as laws or money in that the realization of their function crucially depends on their physical structure. They are also different from physical or natural objects because they are intentionally produced and used by human beings to realize certain goals.' According to their 'dual approach', technical objects are sets of manufactured components that are arranged to create a function. In turn, humans draw upon these functions to complete particular tasks. This dual definition is explicitly and implicitly repeated throughout various accounts of technology in philosophy, social science and the humanities more broadly. As Ihde (2004: 91–108) argues, within the history of twentieth-century criticism, the majority of commentators – from Marx to Heidegger to Adorno – position technology as sets of objects manufactured and used by humans to organize society. For Marx, technology becomes a cornerstone of capital accumulation, which works to 'turn everything into an exchange value for something else' (Wendling 2009: 208). For Heidegger (1982), technology works to enframe the modern world, revealing the environment as a set of resources ripe for exploitation. For Adorno and Horkheimer (1997), media technology is a key tool of capitalism used to pacify the masses and so on. In these cases, what really matters about technology is not the objects themselves, but the effect they have by the way they organize, or are organized by human practices (Kyrre et al. 2008).

From this perspective, technical objects are either reduced to inert clods of matter that have no agency in and of themselves, or understood primarily through the ways they are used by humans. In a world of partly automated digital systems, the idea that smart objects are inert or have no agency is clearly untenable. For instance, Dodge, Kitchin and Zook (Dodge and Kitchin 2005, 2007; Kitchin 2014a; Dodge, Kitchin and Zook 2009) argue that big data systems, which underlie and support smart objects now manage everything from traffic in major cities to airport security and weather forecasting. As Beer (2009: 988) puts it, we exist among 'a vast set of automated communications that are a part of how we live but are often not a part of our day-to-day conscious existence'. Other writers, including Hayles (1999, 2012), Hansen (2000b, 2006a, 2015) and Durham Peters (2015), have pointed out that focusing purely on human experiences of technology and media simply misses too much of what modern technological objects do outside of the phenomenal realm of human perception. In doing so, the kinds of inter-object relation mentioned above tend to disappear from analysis.

The issue becomes how to account for or conceptualize these inter-object relations. Software (Fuller 2003; Manovich 2013) and platform (Bogost and Montfort 2007) studies suggest that in order to engage in detailed cultural and social criticism of digital media, researchers need to understand the mechanisms that make these devices work. Whereas

> much work published in the area of new media largely adopts an Information and Communications Technology model (the shunting of 'content' from point A to point B) for its understanding of phenomena such as the internet or even games, and aims its critical faculties at what happens around or through software … [software studies] … differs by, among

> other things, emphasizing the neglected aspect of computation which involves the possibilities of virtuality, simulation, abstraction, feedback, and autonomous processes. (Fuller 2008: 4)

This body of work suggests that cultural and social readings of technology, in their attempt to correct the determinism of early approaches to media, actually over-emphasize the social and cultural at the expense of understanding the specificity of the technical. But, distinct from determinist approaches to media and technology (e.g. in relation to Marxism see Bimber 1990; Smith and Marx 1994), software and platform studies recognize that examining the technical specificity of digital media is not about attempting to uncover some ultimate truth of its effects. As Fuller (2008: 5) writes in relation to software, 'The purpose of … [a] … lexicon … [of software] … then is not to stage some revelation of a supposed hidden technical truth of software, to unmask its esoteric reality, but to see what it is, what it does and what it can be coupled with'. However, in turning to the 'neglected aspect of computation' (Fuller 2008: 4), there is still a danger that work within software and platform studies reduces technical objects to the logic and materiality of their design and manufacture. From this position, the effects of digital, software-driven smart objects could be read from their material design and functionality.

Other approaches to smart objects move in quite the opposite direction. Rather than focus on the technical specification of smart objects, they tend to anthropomorphize smart objects by ascribing them human features. For instance, Verbeek suggests that smart objects are defined by five key features. Although lengthy, his account is worth reproducing:

> Firstly there is the layer of embedding: ambient intelligence is embedded in the environment, both in the physical sense, hidden in walls, clothing and packaging, and in the social sense: it is possible to communicate with it in a 'natural' way, for example by means of movement or speech. Secondly, this technology is aware of its environment: it responds to what happens

> around it, by detecting movement, reading RFID [Radio Frequency Identification Device] chips, recognizing speech and so on. … The third layer is that of personalization. Ambient intelligence can draw up or retrieve a person's profile and set up interaction with technology that is tailor-made to suit the person in question. An example is a refrigerator that forms a picture of someone's eating pattern and subsequently makes suggestions for the shopping list. … Fourthly, this technology has the capacity to adjust; it not only detects its environment, but adjusts to it in a personalized manner. The intelligent refrigerator in the example might quite well be able to harmonize the menu suggestions to the time of year. The fifth and last layer in ambient intelligence is that of anticipation. This technology can not only respond to its environment, but can also think ahead, like a car that can anticipate the movements of other road users and adapt its speed automatically. (Verbeek 2009: 233)

It is worth reflecting on the language that Verbeek uses to differentiate smart objects from technical objects more broadly. Smart objects are 'aware'; they 'form a picture of people's habits'; they are 'intelligent'; they can 'anticipate' and so on. This language works to personify smart objects, presenting them as having similar or equivalent capacities to human beings. While helpful as a short hand, this language also makes it harder to focus on precisely what distinguishes smart objects from other types of technology or humans. According to Ihde (2005), defining technologies through very general categories or through an equivocation with human capacities can lead to a 'dsytopian totalist' approach, where technology becomes labelled as simply 'good' or 'bad'. Understood in binary terms, it then becomes difficult to develop concrete responses to particular technical objects outside of outright celebration or denunciation. Following Ihde's (2005: 106) conviction that 'there are needs for analyses which can be applied and finely tuned enough to differentiate between types of technologies', I argue that what smart objects do exceeds their computational capacities and how they are used by humans, but at the same time, smart objects should not be described using anthropomorphic language that draws a parallel between human intelligence and technical devices. Understanding smart objects requires that we take technical things as a category of objects seriously and identify what makes them distinct. To do this, we can adopt what Simondon (2017: 19) terms a mechanological analysis of smart objects.

Mechanology refers to the study of technical things as composed of distinct components that form parts of concrete interdependent objects, which have a genesis and development that exceeds that of their human makers. In other words, mechanology is an analysis that asks us to identify the key features that enable a technical object to exist and do work in the world. A mechanologist can then use these features in order to categorize different types of technical objects. As Iliadas (2015: 133–4) usefully points out, these 'categories … are not metaphorical but analogical; that is, they are meant to explain technological similarity via realism about structural relations'. Once we have identified the real structural features of technology, through an analysis of particular components, we can identify what differentiates smart objects from other forms of technical object using the same criteria. As we shall see, following the impulse of mechanology results in a definition of technical objects as allopoietic, homeostatic units that perturb one another. Developing notions of allopoiesis, homeostasis and perturbation is important to demonstrate that what technical objects do cannot adequately be described or understood through an examination of their technical specifications (as in software or platform studies) or through how they are used in relation to human practices (as in Phenomenology or Marxism) alone.

With a definition of technical objects established, we can define smart objects specifically as having an intentionality and protentionality that is shaped by the perturbations that are partly enabled (but never determined) by their allopoietic and homeostatic structure. Analysing smart objects in terms of their intentionality and protentionality, we can begin to see that they do not simply change, alter or over-code an existing space and time.

Rather, via their perturbations, smart objects disclose qualities that generate phases of spatial and temporal intelligibility that cannot be reduced to their technical, cultural or social genesis. Smart objects are thus different in degree rather than kind from other technical objects, but understanding these differences is crucial to developing an account of smart objects as producing phases.

To make these claims the rest of the chapter forms three main sections. The section 'Technical Objects' defines technical objects in general as allopoietic, homeostatic units that perturb one another. With a definition of technical objects established, the section 'Smart Objects' defines smart objects in terms of their intentionality and protentionality. Focusing particularly on the Apple Lightning connector and Siri voice assistant as examples, I argue smart objects' intentionality and protentionality gives them the capacity to generate phases that technical objects lack, in general. The concluding section 'A Quintuplet Model of Smart Objects' reflects on the conceptual distinction between technical objects in general and smart objects in particular and emphasizes the similarities and differences between these objects.

Technical objects

a) Allopoiesis

All technical objects, from lawnmowers, to books, to refrigerators are allopoietic. As Neves (2001: 255) suggests, the term allopoiesis is 'derived etymologically from the Greek words *a΄llos* ("other") and *poiesis* ("creation", "production"), [and] … designates the (re)production of a system through the criteria, programmes, and codes of its environment'. In other words, allopoietic objects are objects produced by a process exterior to themselves.

Allopoiesis is often discussed in relation to inorganic objects, such as a pebble produced through a process of weathering. Within the systems theory of Maturana and Varela (1991) and Luhmann (1986, 1995), allopoiesis is contrasted to processes of autopoiesis, in which a system creates or regenerates the components from which that system is made (such as a biological cell). In relation to technical objects, allopoiesis refers to the human processes that create an object, such as its design and manufacture. But at the same time, the concept of allopoiesis also questions the notion that technical objects are reducible to the processes of human intentionality associated with the design and manufacture of these objects.

From an allopoietic perspective, humans are not masters of technology but intermediaries that attempt to cobble together and manage the indeterminate relations between an object's components in the most efficient way possible. As Simondon (2017: 18) argues, humans are beings 'who regulate ... the margin of indeterminacy ... [of technical objects] ... in order to adapt ... [them] ... to the best possible exchange of information'. For Simondon, information refers to an organized pattern of transference of matter or energy from one object to another, which affects that object in some way. Here the relations between the components of technical objects are always indeterminate, regardless of how cleverly they are designed or how well they are made. Rather than being dreamt up in the mind of a genius inventor and then realized in matter, technical objects emerge from a process of innovation or what Simondon (2017: 26) calls a genesis, where the object unifies itself internally according to a principle of inner resonance. Technical genesis refers to the way that 'technical being evolves through convergence and self-adaptation to itself' (Simondon 2017: 26). To explain this abstract term, Simondon draws upon the example of a petrol engine. Here

> the gasoline engine is not this or that engine given in time or space ... there is a succession, a continuity that runs through the first engines to those we

> currently know and which are still evolving. As such, as in a phylogenetic lineage, a definite stage of evolution contains dynamic structures and schemas within itself that partake in the principle stages of an evolution of forms. (Simondon 2017: 26)

The genesis of the petrol engine is not created or led by humans, in the sense they design increasingly efficient or powerful engines for example. Rather, this genesis emerges from the interior of the technical object itself as successive designers and manufacturers integrate previously separate or abstracted systems of the engine into more concrete arrangements. For example, Simondon (2017: 30) suggests that 'an air-cooled engine is more concrete than a water-cooled engine', because the air used to cool the engine is drawn into the engine by the very working of the engine itself and involves 'thermal infrared radiation and convection … that cannot but take place', whereas a water-cooled engine requires a water pump, which receives its energy from the engine indirectly, via a drive belt. The concrete machine is thus a machine in which each part is closely integrated with one another, often requiring one component to feed or fuel another part. If one part breaks down, then the machine itself quickly stops working. As Simondon (2017: 28) suggests, in concrete machines, there is a 'convergence of functions into a structural unit, rather than ... seeking a compromise between conflicting requirements.' The designers and manufacturers of technical objects can only ever attempt to create more concrete machines in an effort to manage the indeterminate nature of their products. The concept of allopoiesis is therefore useful to emphasize the autonomous nature of technical objects, which emerge from two distinct directions. On the one hand, technical objects cannot be reduced to their design by a particular company or individual because their concretization is always enabled according to logics that precede those companies or individuals. On the other hand, technical objects' allopoiesis means they can operate as autonomous things in the world after they have been produced. This suggests that the designer or creator of an object cannot determine or anticipate what that object can or will do once it has left the factory or workshop. To better understand the autonomy of technical objects when they exist in the wild, we can turn to the concept of homeostasis.

b) Homeostasis

Homeostasis can be defined as a 'relatively stable state of equilibrium or a tendency toward such a state between the different but interdependent element or groups of elements of an organism, population, or group' (Webster 2014 n.p.). Simondon (2009) argues that all technical objects have to exhibit some form of self-regulation in order to exist. In the case of a combustion engine, if the combustion chamber was not sealed then the burning fuel could spread to other parts of the engine, causing it to break down. Simondon suggests that processes of biological and technical homeostasis are similar, but ultimately different from one another. As Simondon (2009: 7) puts it:

> the living conserves within itself a permanent activity of individuation. It is not only the result of individuation, like in the case of the crystal or the molecule, but it is the theater of individuation: not all of the activity of the living is concentrated at its limit, such as with the physical individual. Within the living itself, there is a more complete regime of internal resonance, one that requires permanent communication and that maintains a metastability that is a condition of life.

While the form and shape of an inorganic object, such as a crystal, is determined by its environment alone, an organic object, such as a human body, can regulate itself according to varying internal systems as well. Organic processes of homeostasis in the human body involve receptors, control centres and effectors, which can dynamically respond to events both within and outside of itself (Cannon 1929). These homeostatic processes can also operate in negative and positive feedback loops, working to minimize change (such as when regulating body temperature) or to increase change (such as blood clotting when the body is wounded). Although many technical objects, such as thermostats, also have sensors and regulatory mechanisms to respond to external events their very potential for homeostasis is partially defined allopoietically, through the object's design, manufacture or concretization. As Simondon (2009: 7) goes on to elaborate,

> There is, in the living, an individuation by the individual and not only a functioning that would be the result of an individuation completed once and for all, as if it had been manufactured; the living resolves problems, not only by adapting itself, that is to say by modifying its relation to the environment (which a machine can do), but by modifying itself, by inventing new internal structures and by completely introducing itself into the axiomatic of vital problems.

A key difference between organic homeostasis and technical homeostasis is, therefore, the (in)capacity of technical things to self-generate new internal regulatory systems and mechanisms. As Simondon (2009: 7) puts it: 'The living is the being that is the result of an initial individuation and that amplifies this individuation; an activity not undertaken by the technical object.' In other words, a technical object's homeostasis is allopoietic. While it can engage in various forms of self-regulation, it can't create new grounds of self-regulation by itself.

Reflecting on the two concepts developed so far, we can argue that technical objects are allopoietic and homeostatic and that their modes of allopoiesis and homeostasis shape what a particular object can do. In the next section I want to suggest how technical objects engage with one another is shaped, but not determined, by a particular object's allopoiesis and homeostasis. Rather than existing as completely available or present entities, technical objects perturb one another and disclose particular qualities through these perturbations.

We can theorize how qualities are disclosed by developing the concept of perturbation in more detail.

c) Perturbation

Perturbation refers to the ability of a technical object to disclose qualities of itself, or another object in some basic way. This is linked to Bryant's (2011: 167) claim that objects are 'selectively open to their environments'. According to Bryant, when one object encounters another a perturbation occurs, which is then translated into different forms of information dependent on the object in question. As such, any perturbation is always selective. To clarify this, Bryant gives the example of a rock. As he puts it, 'while rocks … are certainly open to sound waves, they are not, as far as I know, open to signifiers' (Bryant 2011: 167). The concept of perturbation results in a novel account of relationality. For example, Bryant discusses a cat sitting on his lap:

> When my cat rubs against me or jumps on my lap these events constitute perturbations for me. However, as a system I translate these perturbations into information, registering them as signs of affection. In response, I pet my cat to show my affection. By contrast, my cats might merely be seeking warmth or marking me with their scent so as to establish territory. The point here is that no identity of shared information need be present for this interaction to take place and maintain itself. My cat and I are perhaps occupied with each other for entirely different reasons, completely unaware that we have different reasons, yet an interaction and communication still takes place. (2011: 154)

Here, as Bryant (2011: 31) puts it, 'Relations between objects are accounted for by the manner in which objects transform perturbations from other objects into information or events that select system-states. These information-events or events that select system-states are, in their turn, among the agencies that preside over the production of local manifestations in objects.' In relation to technical objects, these local manifestations can be defined as qualities.

According to Graham Harman, a quality is a 'sensual note' (2005: 221) or 'accidental surface profile' (2009: 203) of an object. A quality might be the sharp zest of a lemon, or its particular shade of yellow. For Harman, these qualities are not just apparent to human sense perception when someone sucks a lemon or looks at it from afar. Rather the lemon will disclose different qualities to other objects depending on who or what is being encountered. While a human might be perturbed by the lemon's colour, an insect might be perturbed by its scent. Objects work to disclose various qualities or sensual notes of other objects depending on how they perturb or are perturbed by these objects.

In relation to technical objects, we might say that making a telephone call on a digital cordless phone involves a human voice perturbing the microphone, which perturbs an antenna, which perturbs copper cables, which perturbs the person that you are talking to on the phone. In each case the perturbations are inherently linked, but not shared by any of the individual components of the phone. Without the air and antenna being perturbed by the radio waves the phone would not perturb me by transmitting sound waves to my ear. Each of these perturbations disclose qualities that are necessary for the phone to inform the user of an incoming call, but each component is perturbed in ways that are completely private to each object. When I pick up the phone to answer, I do not experience sensual notes from the radio waves or the antenna, but at the same time the radio waves and antenna are not perturbed by me directly. Instead I experience particular qualities of these objects, such as the smoothness of the plastic body of the telephone in my hand or the high-pitched tininess of the speaker inside the phone's body in my ear.

This notion of perturbation is useful as it allows us to understand how technical objects can have effects on other objects without reducing these effects to the allopoietic or homeostatic structure of the object alone. Technical objects always have the capacity to exceed their allopoietic and homeostatic structure because every relation between objects involves the disclosure of qualities that cannot be fully known or anticipated in advance. In other words, knowing a technical object does not consist in enumerating a list of its technical features or mechanisms, but consists in exploring the qualities that can be disclosed as that object is perturbed by different things. This means that the being of a technical object is 'not the sum of its qualities, but rather qualities' are disclosed by the perturbations of an object (Bryant 2011: 90). From this perspective, technical objects are essentially bottomless. As Bryant (2011: 90) discusses in relation to a blue mug, 'we can say that in principle the mug is potentially an infinite number of colours because there is no limit to the ... relations into which the mug can enter. Consequently, we cannot say that we would finally get the true being of the mug by adding up all ... [of its] ... qualities.' The capacity of technical objects to do things emerges from the tension between an object's allopoietic and homeostatic structure and the contingency of its perturbations, which are ultimately limitless, but partly shaped by which component of an object perturbs a component of another object. In relation to technical objects, we can turn to the example of a simple corded telephone to illustrate this claim.

A corded telephone's capacity to make calls is shaped by the disclosure of a set of qualities, such as the specific electrical signals that are disclosed when the microphone in the speaker is perturbed by a human voice, which perturbs the copper wires inside the telephone cord. This perturbation is shaped by the allopoietic and homeostatic design and structure of the microphone and copper wire, but these qualities are only disclosed when the phone is picked up and a call is made. However, if we were to use the cord from the phone as a bracelet or a trip wire it would disclose very different qualities, which would alter its ability to perturb other things. As a bracelet the cord might disclose qualities such as smoothness, which would perturb the wearer through a sense of comfort or reassurance. As a tripwire the cord might disclose qualities of tautness or tensile strength that might perturb a human by causing them to fall and injure themselves. In each case, different qualities of the cord are disclosed and shape how it can perturb other objects in ways that are not apparent when using the cord as part of the phone to make a call. In these hypothetical cases the particular qualities of the cord are not actualized from a latency or potentiality of the cord that is the product of the event between two entities. Rather, all the qualities of the cord precede any event or encounter and are carried internally in the object. As such, we can state that technical objects such as a corded phone have a limitless number of qualities, partly due to the allopoietic and homeostatic structure of the object and partly because different qualities of the object will be disclosed depending on how it perturbs and is perturbed by other things. No technical 'object is reducible to the events in which it participates' (Harman 2010a: 65) because, while its allopoietic and homeostatic structure shape the kinds of qualities that are disclosed when it perturbs or is perturbed by other things, this structure can never fully determine what these qualities will be.

Smart objects

The three concepts developed above are necessarily broad, providing a general framework for understanding all manner of technical objects, from stone flints, to sunglasses, to microphones, to digital computers. While smart objects undoubtedly form part of a more general category of technical objects, they also have at least two features that differentiate these objects from other technologies, which I term their intentionality and protentionality. It is important to note that in identifying and discussing intentionality and protentionality, I am not suggesting there is a clean or clear distinction between smart and non-smart technical objects. As Simondon (2017: 26) suggests, all technical objects have a genesis, where the beginnings of one feature can be identified in another object. In this sense, proto forms of intentionality and protentionality can certainly be found in non-smart objects. Theorizing smart objects in terms of their intentionality and protentionality is not a matter of claiming or announcing an epochal shift in technical development. Rather the aim is to develop a conceptual vocabulary that is sensitive to smart objects in order to understand how they produce phases.

a) Intentionality

Like all technical objects, smart objects have an allopoietic and homeostatic structure. Smart objects utilize a range of sensor components as part of this structure. For Gabrys (2015: 8) sensors can be defined as 'devices that typically translate chemical and mechanical stimuli such as light, temperature, gas concentration, speed, and vibration across analogue and digital sensors into electrical resistors and voltage signals. Voltage signals further trigger digital circuits to output a series of conversions into zeros and one, which are processed to form readable measurement and data.' Many smart objects, such as smartphones, smart security systems, smart thermostats and autonomous vehicles utilize multiple sensors such as cameras, accelerometers, microphones, radar and sonar among others. The number and combination of sensor components provides smart objects with differing levels of intentionality that is key to both what smart objects can do and how they differ from non-smart objects.

Within phenomenology, intentionality refers to a basic observation that human consciousness is always conscious of something. We do not see light rays hitting our retina; we see a room, or a cat, or a screen. Phenomenologists from Franz Brentano to Edmund Husserl have argued that intentionality is partly what separates conscious from non-conscious beings (Gallagher and Zahavi 2012: 80). As Shaviro (2014: 80) summarizes, 'The commonly held doctrine … is that intentionality is an exclusive mark of the mental or psychological: indeed, intentionality is generally held to provide the definitive principle of a' 'demarcation between the psychic and the physical' (Molnar 2007: 61). Challenging this view, Shaviro suggests that the concept of intentionality can also be extended to inanimate beings, which he refers to as a physical intentionality. By this Shaviro (2014: 81 citing Molnar 2007: 63) means that 'actually existing physical powers "also have that direction toward something outside of themselves that is typical of psychological attributes".' In other words, 'intentionality becomes an implicit striving toward, or a potential for becoming, within the world' (Shaviro 2014: 81). Heil (2004: 440) gives the example of a grain of salt dissolving in water to understand this: 'A power is always of or for some definite kind of manifestation … . The manifestation of powers is most often a reciprocal affair: when a grain of salt dissolves in a quantity of water the dissolving is a "mutual manifestation of reciprocal disposition partners". The relation is symmetrical. … Salt … "points toward" dissolving in water; water "points toward" dissolving salt.'

Based on this observation, the allopoietic and homeostatic structure of smart objects such as the iPhone, which has various sensor components, utilizes this physical intentionality in order to exhibit different modes of smart intentionality. Unlike physical intentionality, which is shaped by the contingency of an object's current state, technical intentionality is enabled and limited by the allopoietic processes of genesis and concretization that produced that particular object. In other words, smart objects could be considered as intentional beings in the sense that they are designed and manufactured to be concrete machines, whose various sensors enable a specific form of 'towardness' to other things, while nonetheless having contingent qualities. It is important to note that these intentionalities are not random, but are usually generated both to furnish smart objects with particular functions and to allow the companies who create these objects to attempt to control how they are used. To unpack the notion of intentionality in more detail we can turn to the very simple example of the Apple Lightning cable.

The Lightning cable is Apple's proprietary charging and data exchange cable that is designed for use with Apple devices, including the iPad and iPhone (henceforth collectively referred to as iDevices). The most recent Lightning cable has been designed to disclose qualities that perturb and are perturbed by iDevices in very specific ways. The Lightning cable, due to its shape, size, number of pins and mode of data transfer will only work with Apple products. More specifically, the cable is reversible, meaning that it can be plugged in with either side facing the front of an iDevice. This is distinct from the older-style Apple charging cable, which is faced, meaning it has a front and a back and the front of the cable has to be plugged into the front side of an iDevice in order to charge it or transfer data. Constructing the Lightning cable and iDevice data port to allow reversible connection did not simply involve the manufacture of a symmetrically shaped cable head. Apple also had to create electronics in the cable that were capable of being perturbed by an iDevice regardless of which way the cable head was inserted. The head of a Lightning cable contains four microchips or 'dies' all of which have various functions: the NXP SP3D2, the STMicroelectronics USB2A, the Texas Instruments TIBQ2025 and a chip from an unknown manufacturer labelled '4S'. Outputs from these chips are connected to eight pins on the tip of the head, which connect to the iDevice when inserted into the data port. When connecting the cable into the iDevice, the iDevice has to 'work out' the orientation of the head. The iDevice can do this because the Texas Instruments chip, which transmits data to the iDevice, is wired into two separate pins, each on different sides of the head. The iDevice only monitors the bottom set of pins, so the iDevice tries to perturb both pins. Depending on which pin the iDevice is perturbed by, the iDevice can then perturb the phone with data and other objects via the cable (for a full technical breakdown of the cable, see Chipworks 2012).

Beyond the physical shape of the Lightning connector and the way the cable communicates to iDevices through their data pins, Apple has added measures to ensure that cables produced by third parties will not work correctly with iDevices, unless the third party pays Apple licensing fees through its MFi (Made For iPhone/iPod/iPad) program. The MFi program enables hardware developers to 'get the hardware connectors and components that are required to manufacture iPod, iPhone, iPad, and AirPlay audio accessories … and access the iPod Accessory protocol specification, the communication protocol used to interact with iPod, iPhone, and iPad' (Apple 2015: n.p.). If a third-party cable does not have the same array of chips as the official cable, it may not work with Apple's devices, regardless of whether the cable uses the correct type of ports or is of the correct shape. In particular, the TIBQ2025 chip uses Texas Instruments' proprietary serial communications protocol, without which the cable cannot successfully perturb the phone. Here, the particular concretization of the Lightning cable's design, which involves the integration of different chips into the core functionality of the cable, works to create a form of intentionality that enables the cable's capacities and increases Apple's control over their devices.

This brief example demonstrates that smart objects are concretized to exhibit a form of intentionality, without confusing this with a kind of human intentionality that arises from the autopoietic feedback loops of consciousness. As the example of the Lightning cable shows, the qualities disclosed by the perturbations of the phone and cable do not have to require interpretation or reflection on behalf of the objects involved. Rather, intentionality can be a more basic form of disclosure that occurs through perturbations that are partially controlled by the sensors, protocols and chips through which these objects disclose qualities in one another.

Indeed, the dual logic of function and control that lies behind the intentionality of the Lightning cable can be evidenced in a range of smart objects. For instance, the Nest Security Cam, which we will discuss in more detail in Chapter 3, utilizes a motion sensor to detect movement and then sends images from the camera to a smartphone. The camera is sold as a convenient way to keep your home safe. However, if you want to retain recorded footage and activate advanced sensing features of the camera, you need to pay for a Nest Aware subscription, which utilizes cloud computing to store and analyse images from the camera. Intentionality should therefore be considered as a basic feature that differentiates smart objects from non-smart objects, but also enables new ways for the manufacturers of these devices to attempt to profit from and control their objects.

b) Protentionality

Depending on their combination of sensor components, many smart objects also have the capacity for protention, which can be understood as the second key difference between smart and non-smart objects. In relation to human perception, protention refers to the capacity to anticipate what is to come next and is linked to broader understandings of temporal consciousness. There are many different theoretical approaches to understanding the human phenomenon of time consciousness. Within new media and technology studies perhaps the most well known of these approaches is the work of phenomenologists Husserl (1991) and Heidegger (1962) as well as Bernard Stiegler's (2009, 2010d) interpretation of Husserl (see for example Hansen 2004, 2006b, 2009). In *The Consciousness of Internal Time*, Husserl (1991) splits human temporal consciousness into three modes: primary retention, secondary retention and protention. Steigler (2010a: 8–9) describes this in the following way:

> Primary retention is that which is formed in the very passage of time, as the course of this time, such that, as a present which passes, it is constituted by the immediate and primordial retention of its own passing. Becoming past, this passage of the present is then constituted as secondary retention, that is, as all those memorial contents which together form the woven threads of our memory.

Very simply, primary retention can be understood as perception and secondary retention can be understood as recollection. Distinct from primary and secondary retention, Stiegler (2010d: 17) defines protention as 'expectation that animates a consciousness'. In other words, protention is how we anticipate what is to come next, based upon our perception and recollection.

Key to Husserl's account of time consciousness is that any human experience has a common horizonal structure. The now is never experienced in isolation but is always intentional in two directions. In one direction the now is experienced in relation to some 'just past' that remains in consciousness and in the other the now is experienced through an 'about to be'. As Gallagher and Zahavi (2012: 86) put it,

> If we look at a pedestrian who is crossing the street, our perception will not be restricted to capturing the durationless now-slice of his movement. Perceptually, it is not as if the pedestrian suddenly appeared out of nowhere; and further we do not have to engage in an explicit act of remembering in order to establish the temporal context of his current position. Nor however, will it be the case that all the previous slices of his movement are perceptually present in the same way as his current position. If that were the case, the pedestrian would perceptually fill the entire space he has just traversed.

From this perspective the temporal structure of human consciousness is permeated with meaning. Gallagher and Zahavi (2012: 86) describe this in the following way: 'What we mean by retention is that at any moment what we perceive is embedded in a temporal horizon. Its meaning is influenced by what went before, which still intentionally registers in our awareness. For the just past tone to be intentionally retained is for its meaning or significance to be retained as just past.' The fact that retention imbues the present with meaning is often used as one way for phenomenologists to distinguish between the capacities of humans and non-human animals (Buchanan 2008; Calarco

2008). The horizonal structure of consciousness gives humans seemingly unique capacities for recollecting the past and the ability to plan for future events, which non-humans apparently lack (Roberts 2002).

However, the intentionality afforded by smart objects' allopoietic and homeostatic structure would seem to challenge such distinctions and suggests that many smart objects are capable of some kind of protention. Take for example a context-aware feature of the iOS9 firmware, 'Hey Siri'. Siri is Apple's Artificial Intelligence personal assistant that forms part of the core functionality of the iOS mobile operating system shipped with iPhones, iPads and Apple Watches. Users can talk to Siri and ask it to fulfil a range of functions, from dictating messages and emails, to asking for directions, to asking for weather forecasts and adding notes to the calendar. Siri's functionality is based on the integration of three main components: conversational interface, personal context awareness and service delegation. Conversational analysis involves recording, analysing and translating human speech into information that Siri can respond to. Siri has personal context awareness in the sense that it can learn language habits and adapt to individual preferences. Service delegation allows Siri to access other apps to respond to requests and gather information that is relevant to the user's question or query. When asking for directions, Siri uses the Apple Maps function; when recording a reminder, Siri adds a note to the Calendar app, and so on. Siri also draws upon the locative capacities of iDevices. As long as location services are enabled, a user can ask for a nearby place to eat lunch and Siri will respond with a list of names of restaurants that are proximate to the user's position.

Hey Siri is a feature that uses the M9 motion tracking coprocessor in the iPhone. Specifically it is always 'listening' for the voice command 'Hey Siri'. When the processor 'hears' this phrase, it activates Siri on the phone. Apple describe Hey Siri as 'always listening'. What this means is that the M9 coprocessor takes regular samples from the microphone in the iPhone. Specifically, the analogue to digital convertor in the phone is perturbed by the vibrating air waves, which discloses qualities of those airwaves as electrical signals, which then perturb the processor. It does this by sampling the sound 'by taking precise measurements of the wave at frequent intervals' (Grabianowski 2006: n.p.). Next, this data is sent to Apple's servers, where statistical analysis of the soundwaves takes place. Specifically:

> the signal is divided into small segments as short as a few hundredths of a second, or even thousandths in the case of plosive consonant sounds – consonant stops produced by obstructing airflow in the vocal tract – like 'p' or 't'. The program then matches these segments to known phonemes in the appropriate language. A phoneme is the smallest element of a language – a representation of the sounds we make and put together to form meaningful expressions. There are roughly 40 phonemes in the English language (different linguists have different opinions on the exact number). (Grabianowski 2006: n.p.)

Once divided into small segments the server examines the phonemes it has identified in relation to the phonemes that surround them. It then analyses the set of phonemes in relation to a statistical model that compares the patterns to existing known words and phrases and uses this comparison to decide what words the user was likely to have said. What is key to the success of this technique of analysis is that the model must have a pre-existing library of words and phrases to compare the sampled sounds to. If it does, then it can be perturbed by the individual sounds and disclose qualities of words that it can respond to.

Although Siri is ultimately a distributed object, spread across a range of devices, servers and software, it would seem to utilize elements of a retentional structure of primary and secondary retention that enables protention. The M9 coprocessor samples sound waves that enter the microphone, which can be understood as a form of primary retention, which are analysed via its secondary retentions. These are the sets of words and phrases that have been previously modelled by the system and retained on Apple's servers. Primary and secondary retentions then enable protention in the sense that Siri uses these retentions to recognize when the phrase 'hey Siri' has been uttered.

Of course, one might argue that Siri does not have time consciousness in the sense that Husserl understood, precisely because Siri's primary retentions are not infused with a secondary retention and protention. For Husserl, strictly speaking, secondary retention is not a form of memory, because it is always present with a primary retention and shapes how that retention is understood (Gallagher and Zahavi 2012). In other words, Siri's secondary retentions (the data stored on its servers) are not actively recalled in the same way a human thinks back upon a memory such as a birthday last year.

But, if we investigate Husserl's account of retention more closely, some interesting analogies arise. Although Husserl was not interested in the underlying neurobiological processes involved in the retentional structure of consciousness, empirical evidence from the cognitive sciences has begun to question the relationship between primary and secondary retention as inherently glued together in the way Husserl suggested (Pöppel and Artin 1988; Varela 1999). For example, one way that the phenomenological model of retention and protention has been linked to cognitive accounts of temporality is through the notion of the human brain and body as a self-organizing dynamic system. Here 'time emerges as parts interact in a non-linear fashion, reciprocally determining each others behavior through a process of self organization in which the parts remain dynamically coordinated to each other over a period of time … . With respect to cognitive experience, different processes in the body, brain and environment become temporally coupled in dynamic coordination' (Gallagher and Zahavi 2012: 89). In other words, different parts of the brain work together in relation to the environment to construct a horizonal structure of temporal consciousness. According to cognitive science, these processes are organized around three temporal scales: 'the elementary scale (varying between 10 and 100 milliseconds), the integration scale (varying from 0.5 to

3 seconds) and the narrative scale (anything longer than this)' (Gallagher and Zahavi 2012: 89).

From a cognitive perspective, the basic 'unit' of retention for humans is the 'minimum amount of time needed for two stimuli to be consciously perceived as non-simultaneous' (Gallagher and Zahavi 2012: 89). Again, as Gallagher and Zahavi (2012: 89) put it, 'In terms of a dynamical systems model, the cell assembly … [in the brain] … must have a relaxation time followed by a bifurcation or phase transition, that is, a time of emergence within which an experience arises, flourishes, and subsides, only to begin another cycle.' In humans, this takes place neurophysiologically at the elementary scale (between 10 and 100 milliseconds) and in lived experience at the integration scale (between 0.5 and 3 seconds). In some ways, Siri and the iPhone could be said to operate around a similar technique of sampling. In relation to the iPhone, the M9 coprocessor also has a minimum or maximal unit of sampling at different scales. For example, the microphone and processor may sample sound waves thousands of times a second and these are then sent to the Apple cloud server, processed and returned at a scale of one to two seconds. In other words, Siri also has a basic unit of primary retention and this has a relation with any secondary retention, dependent on whether the sound waves are being analysed locally by the phone, or sent to Apple servers for analysis. Indeed, from a cognitive science perspective, even a human retentional continuum, as described in conscious experience, is actually constituted by more or less discrete units of retention. As neurobiology shows, these retentions only appear as a continuum because of the speed that they are processed on a neurophysiological level (Pöppel and Artin 1988). This means that while human temporal consciousness appears to be a continuous qualitative process on the register of perceptual awareness, it is actually discontinuous on a biological register.

Based upon this similar technique of sampling, it would be possible to state that smart objects such as the iPhone have a protentionality. The protentionality particular to smart objects can be defined as their ability to await specific forces or signals in the environment and respond to these forces and signals through the way they perturb or are perturbed by other things.[1] Smart object protentionality is different from human temporal consciousness in at least three ways. First, for smart objects, the meaning or significance of retentions and protentions are not organized around earthly desire, cultural appropriateness or biological need, but by the allopoietic and homeostatic structure of these objects, which shape the way they can perturb and be perturbed by other objects. Second, the discrete perturbations from a smart object's sensors, such as a microphone or camera, that enable retention and protention, can't be joined up to enable the disclosure of a world as a set of holistic significances and meanings. Put another way, the discreteness of a smart object's retentions can't be added together to give that object a coherent picture, image or understanding of an environment because the qualities of objects that the perturbations of smart objects disclose are always partial and task specific. While humans can anticipate what is to come next because they draw together a series of perceptions and memories in order to orientate experience towards the future, smart objects can only await other objects that their sensors are capable of perturbing and being perturbed by. Third, as I will explain in greater detail in Chapter 4, while a useful placeholder for now, the term retention is not that appropriate to understand how smart objects hold or contain experience. Rather, the retentions that enable protention are better understood as the durability of perturbations within the components of smart objects.

[1]This definition should not be confused with what Hui (2016) has recently termed the 'tertiary protention' of digital technologies. For Hui (2016: 221–2), tertiary protention refers to the way digital technology 'becomes a significant function of the imagination' whereby 'orientation becomes more and more an algorithmic process that analyses and produces relations to pave the way for the experience of the next now or immediate future'. Distinct from Hui's focus on the effects of technology on human protention, my discussion of smart objects focuses on the way that smart objects themselves have some capacity for protention and how this is different in kind from human protention.

Just like intentionality, protentionality is not limited to the Hey Siri functionality of the iPhone. Beyond voice recognition, as we shall see in Chapters 4 and 7, smart vacuum cleaners such as the Dyson 360 Eye robot and smart vehicles such as the Google car and Tesla Model S utilize radar, camera, microphones and other sensors to generate different forms of protentionality that is key to what they can do. Like the intentional features of smart objects, the protentionality of objects such as Siri is also sold to consumers on the basis of ease and convenience. But, at the same time protentionality can be utilized in an attempt to capture, control and sell users' data to increase the profitability of these objects. For instance, microphone sensors on a range of smart objects such as LG and Samsung Smart TVs, Xbox Kinect and Amazon Echo can sample consumer speech and potentially sell this data on to third parties or use it to provide targeted advertising through these devices (Adams 2013).

A quintuplet model of smart objects

This chapter has developed a quintuplet model of smart objects. Like technical objects in general, smart objects are allopoietically produced units that exhibit a degree of homeostasis and perturb one another. But different (in degree) from other technical things, smart objects also have an intentionality and protentionality. This intentionality and protentionality allows smart objects to purposively and selectively perturb and be perturbed by other objects and respond to these perturbations with perturbations of their own. Distinct from the dual approach outlined by Kroes and Meijer (2006) in the opening section, this quintuplet model emphasizes inter-object relations that do not necessarily appear to humans as key to defining and understanding what smart objects are and what they do.

Understanding smart objects through the concepts of allopoiesis, homeostasis, perturbation, intentionality and protentionality means beginning with the objects that appear within a given situation and investigating all the different ways these objects might perturb or be perturbed by one another to disclose or express different qualities, while recognizing that any individual smart object's qualities can never be fully known or resolved. Smart objects can exist alongside one another without entering into direct contact or relation. But at the same time, this does not mean a smart object and its capacities are random or can suddenly change without reason or warning. As we have seen, smart objects can be designed, manufactured, distributed and organized in such a way to perturb and disclose more or less repeatable qualities of other objects.

With this point in mind it is important to reiterate that the intentionalities and protentionalities of smart objects are not innocent or accidental, but are organized by a commercial impulse. Specifically, intentionality and protentionality are used to sell smart objects that promise convenience and ease of use, while also enabling new forms of control by the manufacturers of these objects. This commercial logic is linked to a series of techniques for modulating phases and the spatio-temporal intelligibilities they attempt to generate. The next two chapters turn to examine how smart objects produce phases and investigate these techniques of modulation. Chapter 3 examines the notion of phase space and Chapter 4 turns to phase time. Any one phase is both spatial and temporal, but splitting the various aspects of phases into distinct chapters serves a useful analytical function. By understanding how phases generate spaces and times using different examples, it will become clear that phases are powerful fields through which life can be organized for both human and non-human beings.

3

Spaces

Writing in 'The Real-Time City?' Kitchin summarizes accounts of smart cities that have become pervasive in popular and academic discourse. Kitchin (2014b: 2) suggests smart cities are usually defined in one of two ways:

> On the one hand, the notion of a 'smart city' refers to the increasing extent to which urban places are composed of 'everyware' (Greenfield); that is, pervasive and ubiquitous computing and digitally instrumented devices built into the very fabric of urban environments (e.g., fixed and wireless telecom networks, digitally controlled utility services and transport infrastructure, sensor and camera networks, building management systems, and so on) that are used to monitor, manage and regulate city flows and processes, often in real-time. … On the other hand, the notion of a 'smart city' is seen to refer more broadly to the development of a knowledge economy within a city-region.

In addition to the distinction that Kitchin makes between technical infrastructures of smart objects and the knowledge economy these objects support, we can also differentiate between accounts of smart cities that focus on ideal or formalized tropes around the kind of smart city that might exist in the future and accounts of the reality of smart cities now. The more abstract or formal notions of the smart city tend to concentrate on a vision of urban space as a fully integrated system or holistic network, where individual sensors or components are connected to a centralized control system that allows cities to be managed as efficiently as possible. For example, Batty et al. (2012: 482) suggest that smart objects constitute a new 'planetary nervous system' creating a situation where 'a city might become smart [or] … sentient … is fast becoming the new reality'. Formalized accounts of smart cities tend to position smart objects as a kind of overlay that sits on top of an existing space or infrastructure, even if the creators of these objects seek to transform that space. For example, Batty et al. (2012: 489) argue that smart objects help manage 'real urban traffic' by enabling urban managers to understand 'how people really make choices and how these choices affect the development and spreading of congestion' of a pre-existing road network. Or, as Graham (2017: 44) suggests in his discussion of augmented reality that can be accessed through smart objects such as smartphones: 'Information can now augment and be tethered to places; it can form parts of the layers or palimpsests of place. A building or a street can now be more than stone, bricks and glass; it is also constructed of information that "hovers over" that place: invisible to the naked eye, but accessible with appropriate technological affordances.'

Alongside accounts of layers and palimpsests, other writers working on smart urbanism have tended to emphasize the disjunctive nature of smart systems as a messy overlapping set of objects that do not form a holistic network. As Gabrys (2015: 247) suggests, while cities might be filled with all manner of sensing objects, 'it is still not entirely clear at what point this proliferation of technologies might cross a threshold to constitute a digital infrastructure or fully formed smart city development'. Gabrys does not understand smart objects as an overlay on the spaces of the city; her account of smart cities points to the ways that smart objects actively alter spaces. From this position, smart objects do not 'merely detect preformed environmental data' (Gabrys 2015: 10). Instead, 'distinct environmental relations, take hold, and are programmed with and through these technologies' (Gabrys 2015: 10).

In contrast to formal, abstract or empirical accounts of smart cities, I argue that smart objects need to be explicitly theorized in relation to space as such. Smart objects don't act as a grid or overlay on top of a non-smart space, but nor do they simply modify that environment. Instead, smart objects can be understood as generating phase spaces, which modulate spatial intelligibility for the humans and non-humans that exist alongside smart objects. To define and unpack the concept of phase space the rest of the chapter moves through the following steps. The section 'Phase Space' defines the concept of phase space by developing Heidegger's thought on tools and space. Reading Heidegger, I suggest that his work on space can be productively reimagined on non-anthropocentric grounds to show how technical objects produce distinctions between near and far, which I integrate with aspects of Simondon's work to develop the term phase. The section 'Modulating Phase Spaces' discusses the concept of modulation and intelligibility in relation to the idea of phase space. The section 'Diffusion, Partition and Envelopment' identifies three logics of phase space modulation, which I term diffusion, partition and envelopment. To make sense of these logics I turn to a number of examples, including: Sun, a smartphone weather app; Nest Cam, a home surveillance camera; and Lumo Run, a smart running tracker. To conclude I point to a range of other smart objects and briefly demonstrate how they also produce phase spaces that modulate spatial intelligibility.

Phase space

Key to understanding the concept of phase space is Heidegger's (1992a) term *ent-fernen*, which is translated into English as remotion. Remotion is the basic form of relation between humans and things that makes an environment appear as a set of objects that are spatially discrete or distinct from one another. For Heidegger (1992b: 191), the relations between objects are 'encountered in references'. The term 'reference' can be defined as a kind of relational usability that humans use to make sense of the world. For example, as I write these sentences I sit in my office and notice my computer, telephone, desk, chair, carpet, paint, white board, printed sheets of paper, a jacket, pens, books, light fittings, a radiator and so on. If I were to count these myriad objects there would likely be hundreds within my immediate vicinity alone. The references these objects have are the ways they might be used and the tasks they would help me achieve. These objects form what Heidegger terms a referential totality because each individual piece of equipment or object requires the other to be used and thus made sense of. Without a plug socket the computer could not be plugged in, without a keyboard I could not enter commands into the computer, without a window or electric bulb I would not be able to see to turn the computer on and so on. In the same way, the computer allows me to check emails and write this document, the chair allows me to sit at my desk, the desk holds the screen at the right height for me to see it and so on.

Heidegger develops his account of references to show that space never appears in a blank or neutral way, but always through a set of objects that organize the experience and perception of the user. This account of references results in a holistic understanding of space: 'My encounter with the room is not such that I first take in one thing after another and put together a manifold of things in order then to see a room. Rather I primarily see a referential totality as closed, from which the individual piece of furniture and what is in the room stand out' (Heidegger 1992b: 187). These sets of references in turn create a sense of space through a process that Heidegger calls remotion.

> [It is] … only on the primary basis of … remotion, that the chair, insofar as it is there in a worldly way, as such is removed from me, as such has a possible nearness or distance to me, only on this account is it possible for the chair to be remote from the window, and that we can designate this referential connection of the two as remotion. … Spacing is ontologically

> founded in remotion and can only be discovered and defined when there is remotion. (Heidegger 1992b: 225)

In other words, space appears through the distinction between objects that is enabled by various references and the uses they imply. For Heidegger, space is not simply a box in which activity takes place. Instead it is relations of nearness and farness that are opened up by the way a human distinguishes between things in order to use them. The process of remotion creates what Heidegger terms regions. A region can be defined as 'within the range of something that lies in that direction' (Heidegger 1962: 136). For example, on hearing a door bell we open the door to see who it is. In doing so, the region of the door, the door frame and the building appear not only as distinct things, but also connected to one another. Through this dual process of connection and separation space appears as the distance between these objects, which both separate and connect us to the person on the other side of the door. Regions are thus the distinct areas that form part of a particular practice or activity.

The holistic nature of Heidegger's account of referential totality, remotion and region is based upon an anthropocentricism. For Heidegger, while technical and non-human objects encounter one another, this is not in the sense that they form references with one another. A simple leather shoe has a reference to human feet and the act of walking, but the components from which the shoe is made do not have a referential relation to each other. In other words, the referential totality between objects primarily appears to humans and is not a feature of the things themselves. Heidegger differentiates between human relations with things and non-human relations with each other by introducing the idea of dependence. In relation to a shoe, Heidegger (1992b: 193) argues, 'The work itself has a way of being dependent on, the shoe of leather, thread, nails, leather from hides taken from animals which are raised by others and so become available.' Dependence, for Heidegger, is then a kind of brute physical encounter, rather than a set of references or significances for the objects themselves.

In the same way that referentiality is constituted by a human experience of a set of objects; for Heidegger remotion is also something particular to human beings rather than an objective process that happens in the world. In his words, 'Remotion does not refer to the spacing between two points, even when we do not take them as pure points but as worldly things (say the distance of the chair from the window). It refers rather to the temporally particular nearness or remotion of the chair or window to me' (Heidegger 1992b: 225).

For Heidegger (1992b: 229) a region can only exist when there is a particular being that has some kind of issue with a particular object and approaches it for some reason: 'a region … can only be discovered as such because disclosive being-in-the-world is itself orientated'. In other words, the recognition of distance is not a feature of the relation between the objects themselves, but something particular to Dasein (Heidegger's term for being-there or human being) that it assigns to other things. As such, from a Heideggerian perspective, the actual objects of the door, doorbell and door frame do not themselves constitute a region alone. It is only when a human responds to the door bell, or opens the door, or awaits a guest that these objects come to have a spatial relation. Despite this point, Heidegger does recognize that the usability of an object and in turn its reference for a human being is partially defined by that object's material structure, which exists regardless of a human's concern for, or perception of, that object. In relation to particular tools he argues:

> The range of usability of a tool is narrower or wider. A hammer has a wider range of usability than a watchmaker's instrument which is tailored precisely to his particular kind of concern. The narrower the sphere of use, the more unequivocal the reference. Within its sphere a tool can be applied now to particular parts and pieces of that which is to be produced and then to particular stages in the course of production. (Heidegger 1992b: 191)

In other words, a standard claw hammer has a number of uses across a range of different contexts such as hammering in nails, removing nails, wrenching open crates or smashing objects to pieces. These uses are themselves partly enabled by the weight and density of its metal head, the length of its handle and so on. In the same way, a tiny watchmaker's screwdriver has a very different range of uses than the hammer. Its thin shaft and short length depict that it could not be used to hammer in nails or would snap if used to try and wrench open a crate. While Heidegger reduces the ability for non-human objects to encounter one another to dependences that cannot form references and thus regions with one another, he also recognizes that the different material forms and capacities of objects do affect the kind of references that can be formed between humans and things. In other words, an object's reference is not purely assigned by a human using that thing but is also partly internal to that object, which shapes how it can be used.

Pushing this point, we could state that smart objects can in fact form references with one another in ways that are not reducible to mere dependences and in turn participate in process of remotion that can create regions. In the case of smart objects, references are not attached to objects by humans, but exist in objects despite the presence of a human being. Smart objects can generate references and participate in processes of remotion due to their intentional and protentional structure that is enabled by a range of sensor components, such as antennas, microphones, accelerometers and so on. For example, a smartphone, such as an iPhone, is not an object that simply sits mute in a room 'depending upon' or waiting for a human to activate it or enrol it into a set of references in relation to human practices or understandings. The iPhone is already intentionally and protentionally perturbing other objects without or despite a human user and creating references between these objects. Some of these references would include the way the iPhone perturbs and is perturbed by sound waves that enter the microphone on the phone when using Siri and how Wi-Fi radio waves perturb the antenna, which allows it to pull data from remotely accessed servers for email, send location information to a variety of services and so on. While not necessarily apparent to the human user, these perturbations do participate in processes of remotion and thus generate regions (understood as the appearance of spatial relations between objects) independently of the human.

To explain this further, think of the same iPhone connected to a Bluetooth speaker that is playing music in a room. The speaker and phone are intentionally perturbing one another through the Bluetooth connection. This process of perturbation is a form of relational usability in the sense that each object is seeking out the other in order to operate. The iPhone requires the Bluetooth signal from the speaker to be receiving on the same frequency as the phone in order to perturb the speaker and disclose the qualities of these frequencies as sound waves. At the same time these perturbations create a region through a process of remotion. As sound from the speaker perturbs the walls and other objects in the room this discloses particular qualities of these objects. Depending on the location of objects and the power of the speaker, sound echoes and diffracts, creating the appearance of objects as distantiated from one another. As Kanngieser (2012: 345) points out, 'Spaces manifest sound, even if the sound energy does not originate from the space itself; this occurs through reverberation and reflection – spaces, through their material densities and gaps, modulate and refract sound in peculiar ways.' For example, how far amplified sound will travel depends on the power and size of the speakers used as well as the objects that make up the environment in which the speaker is placed. Perhaps a piece of paper sits on the same table as the speaker. If the speaker is at full volume, the vibrations from the speaker may cause the piece of paper to drop and fall to the floor. Through this simple action, spatial relations of near and far between the speaker, table, paper and phone both appear and change. In other words, the way the phone and speaker perturb one another contribute to a process of remotion through which space appears as a set of distinct objects, regardless of the presence of humans.

In many cases the remotions that smart objects contribute to, and the relations of near and far they disclose, only become apparent to humans when the perturbations between smart objects break down. Think of losing a data signal and being unable to access the internet on a smartphone when entering a tunnel on a train. Here the references between the phone and the network tower or satellite have broken down precisely because the objects involved can no longer perturb one another. In turn, these perturbations shape how relations between near and far appear for the human using the phone. For example, someone sitting on the train engrossed in a book may not notice the length of the tunnel. However, for the person trying to access the internet, the length of the tunnel the train is travelling under suddenly becomes very apparent as they wait to reach the other side in order to regain a signal and continue their browsing. In both these examples, the 'regions' smart objects generate are not a matter of objects attaching or transferring meaning to or from one another. Instead these regions contribute to how space appears by disclosing particular qualities of objects to generate relations of nearness and farness.

To avoid the anthropocentric associations that are linked to Heidegger's discussion of remotion and region, we can suggest that the references smart objects form and how they participate in a process of remotion are better named as phases. For Simondon (2017: 173), the term 'phase' refers to how objects emerge and change within a milieu. Bardin (2015: 36–7) gives the example of a super-saturated solution to make sense of this: 'In a saturated solution one can speak of different phases (solid and liquid, for instance) which can succeed one another, but can also co-exist, and, under certain system conditions, it is the random presence of a seed crystal which can determine the passage (even partial) from one phase to another' (also see Lapworth 2016). Repurposing the word, I use the term phase specifically to emphasize the temporary yet successive and overlapping nature of the spaces smart objects generate. Here, the concept of phase 'takes on a double meaning. ... On the one hand it refers to a succession of states, a process which at times gives the idea of an evolution from … [one space to another]. … On the other hand … the term … refers to the simultaneous presence of multiple … [spaces] – not necessarily harmonised' (Bardin 2015: 37). With this double meaning in mind, phase spaces can be defined as the relations of proximity and distance produced by the qualities disclosed by smart objects as they intentionally and protentionally perturb and are perturbed by other smart objects and objects in general.[1] Practically, the term phase is useful to differentiate between the way humans experience space through assigning references to objects (remotion) and the way that smart objects perturb each other to create regions that exist alongside, outside of or despite human experience or awareness (that I term phase spaces). This definition of phase space has a number of implications for thinking about space in relation to smart objects. For now, at least three of these can be explicitly identified.

First, the term 'phase' highlights the way phase spaces can be both multiple and overlap one another without necessarily coming into contact. When we recognize that smart objects perturb one another alongside or outside the purview of human experience, it quickly becomes clear that the variety of smart, technical and other non-technical objects present at any one moment can create multiple phase spaces. This is because a single perturbation produced by a single smart object such as a smartphone can disclose different qualities of multiple other objects and, thus, contribute to the production of more than one phase space. Returning to the iPhone and Bluetooth speaker example discussed earlier, we can identify at least three overlapping phase spaces. The

[1]My concept of phase space should not be confused with the geographer Martin Jones' development of the term. For Jones (2009: 498–9), 'Phase space is an abstract concept that describes the dynamics and geometry of systems with non-constant parameters being performed within four dimensional space-time. Phase space captures all the possible spaces in which a spatiotemporal system might exist in theoretical terms.' Jones develops the concept of phase space to respond to what he sees as overly relational accounts of space that have difficulty dealing with 'boundedness, inertia and power'. Distinct from this, in my account, phases are the regions opened up by the way smart objects disclose qualities in other objects, which enable space to appear to the humans and non-humans that find themselves alongside these phases.

first phase space would be the perturbations between the iPhone and a wireless router. Perhaps the iPhone is using Spotify to broadcast music. Here the iPhone would need to perturb and be perturbed by the router in order to disclose qualities of the music as sound waves for the speaker. In this case relations of near might appear when the Wi-Fi icon is active on the phone and far would appear when the phone could not be perturbed by the signal from the router. The second phase space would be generated by the perturbations between the smartphone and the wireless speaker. Here the relationship between near and far between the phone and speaker is organized by the power of the Bluetooth signal. When the Bluetooth signal from the phone cannot perturb the speaker and disclose qualities of the MP3 as sound, then the spatial relations between phone and speaker appear as far. But when the phone can perturb the speaker, spatial relations between phone and speaker appear closer or 'in range'. The third phase space might be the spatial relations that are disclosed between the sound from the speaker and a human listening to the music. Here near and far could be organized around the volume from the speaker and the sensitivity of the listener's ear. Near is when music being played through the speaker can be heard clearly and far is when the music is indistinct or quiet. In turn, these relations of near and far might also be influenced by other non-smart objects. For example, the thickness of the walls where the speaker is being used or the type of flooring material, such as carpet or wood, can all influence how sound travels and in turn shape what appears near and far for the human listener. What is crucial here is that all three phase spaces and the distinct relations of near and far they disclose are necessary for the human on the scene to hear the music, but for the most part the first and second phase spaces are not experienced by the human listener at all.

With this example in mind, it is clear that phase spaces are not moments in a process that add up to a singular or totalizable space, but a set of overlapping and often disjunctive sets of perturbation that create relations of near and far that are localized to particular objects. Following Nancy and Barrau (2015: 2), an account of phase spaces asks us to recognize 'that there is no longer any room to speak about … being "in" the world, like contents in a container … . Instead, we must learn the unique yet nonunified, universal yet multiversal existence of all ensembles and everything together.' In short as we have shown with the example of the Bluetooth speaker, objects 'can in fact exist according to several phases that are present at the same time' (De Boever 2012: 221).

Second, the concept of phase emphasizes that the spaces of smart objects are not the metrical distance between two points, but the process of generating the appearance of spatial distinction or differentiation between objects. In turn, phase spaces shape how humans assign references to objects, which enables space to be measured and understood metrically and so appear as a distinct or detached container within which objects, references and relations between near and far are located. From this position, remoteness or distance is the basic condition of object relations. As Heidegger (1992b: 227) puts it, 'It is because world in its very sense is "remote" that there is something like nearness as a mode of distance.' While phase spaces are not geometrical, geometrical understandings of space do not disappear in such an account. As Heidegger argues, geometry is not incorrect or false. But it is a secondary way in which space is experienced. In his own words, 'spacing is a deficient remotion' (Heidegger 1992b: 228). This means geometry, as a form of measurement, can only take place after something like a process of remotion has occurred. In relation to smart objects, it is only once objects perturb and disclose qualities of one another that quantitative spatial distance between them can be measured. For instance, in the previous example of the Bluetooth speaker it is only once the Wi-Fi icon on the iPhone screen turns blank or the music stops playing from the speaker that it is possible to take a tape measure and record the distance at which the phone is 'out of range' and thus 'far' from the router (perhaps 20 metres).

Third, the concept of phase space offers an asymmetrical account of object relations and a nonlinear version of spatial causality. As smart objects and other things only ever disclose a small number of their many qualities, one can never be sure of what a phase space can do, because it is contingent on how smart objects perturb and disclose qualities in one another and other non-smart things. This is an important point to recognize in order to avoid what LaMarre calls an adaptionist account of the capacity of phases to change. For LaMarre (2013), adaptionism refers to a system or object's ability to adapt to conditions that exist outside of it. The problem with overly adaptionist models of change is that they 'place too much emphasis on environmental pressures on the one hand, which leads, on the other hand, to the notion that, over time, as environments change, organisms are stuck with the adaptions produced by prior environments' (LaMarre 2013: 101). Opposed to this adaptionist account, the concept of phase recognizes that while external factors (e.g. non-smart objects that perturb and are perturbed by smart objects) are important, 'what happens inside the machine is equally important' (LaMarre 2013: 101). Phases change through processes of what Simondon (2017: 62) calls recurrent causality. LaMarre (2013: 101) describes recurrent causality in the following way: 'The technical individual effects internal changes and simultaneously generates a recurrent rapport with its external milieu, which allows it to interact with the world and to produce a transformative series.' As such, the causality of phases becomes a matter of how 'specific technical bodies have the power to' (LaMarre 2013: 101) perturb and be perturbed by components internal to their structure and how the space-times these perturbations disclose feedback into the ways the same objects can perturb one another.

Modulating phase spaces

The construction of phase spaces and the disclosure of relations of nearness and farness are not just accidental. Smart objects are regularly designed and created to modulate space-times in order to alter the spatio-temporal intelligibility of the humans and non-humans who find themselves perturbed by these smart objects. I use the term intelligibility here, rather than consciousness or perception, to emphasize that phase spaces don't just modulate human consciousness of a space. Rather, smart objects modulate the appearance of space itself. Instead of an anthropocentric term that refers to how an object appears to consciousness, or as a false or partial facsimile of an object's true form, from a phase perspective appearance refers to an object's 'coming into presence, presenting itself' (Nancy and Barrau 2015: 53). This is because phase spaces do not emerge as the relations between objects, but are appearances that are disclosed by objects as they perturb and are perturbed by other things.

A turn towards thinking the phase spaces of smart objects therefore calls into question accounts that focus on humans' cognitive and sensory capacities as the key site for theorizing the effects that technical objects have on humans. For instance, Hansen (2015: 65) argues that 'despite its qualified displacement in our twenty-first-century media ecology, consciousness remains a crucial – perhaps the crucial – factor for understanding the singularity of human experience as it is being transformed by contemporary media networks'. Here, consciousness is understood to be a target of modulation, where modulation is defined as a 'propensity, where the function of media would be less that of providing stable objects for consciousness, or even stable sedimentations of memory, than that of continuously tweaking on going, largely precognitive tendencies in light of desired experiential outcomes' (Hansen 2015: 214).

As we shall see in the following sections, from a phase space position, smart objects do attempt to modulate human behaviour according to a set of commercial logics. However, modulation is not just a matter of manipulating the structures of human consciousness or unconsciousness. Here 'machines do not act on consciousness as such' (Amin and Thrift 2017: 55), but neither do they 'impact directly on the affective level' (Amin and Thrift 2017: 55) to create various levels of 'technological unconsciousness' (Thrift 2004) either.

Instead, modulation can be defined as a kind of 'self-deforming cast that will continuously change from one moment to the other, or like a sieve whose mesh will transmute from point to point' (Deleuze 1992: 4). Modulation in relation to smart objects operates through the way smart objects literally alter relations between near and far as they intentionally and protentionally perturb objects to disclose qualities that create the sense of near and far. This nearness and farness is experienced not just by humans, but also by other non-humans that are exposed to the qualities disclosed via smart objects and through which space becomes intelligible to them as well. To understand a variety of these logics of modulation in action, we can turn to a number of different smart objects, namely: Sun, a smart phone weather app; Nest Cam, a smart security camera; and Lumo Run, a smart running tracker.

Diffusion, partition and envelopment

Sun, Nest Cam and Lumo Run are good examples of how smart objects generate phase spaces that disclose relations between near and far to modulate the spatio-temporal intelligibility of humans who use these objects. Each object operates according to a different logic of modulation that shapes spatial intelligibility of the environment in different ways. I term these logics diffusion, partition and envelopment. Unpacking these multiple logics, it will become clear that the modulation of spatial (and temporal) intelligibility cannot be explained through processes of reduction, abstraction and quantification alone. Smart objects do not modulate intelligibility by 'subordinat[ing] qualitative … [lived experience] … to the quantitative … [realm of numbers and calculation]' (Moore and Robinson 2016: 2787). Rather, this modulation operates by generating qualitative shifts in the appearance of things that cannot be reduced to quantitative or probabilistic states.

a) Diffusion

The first logic of modulation could be termed diffusion. Diffusion refers to how phase spaces can muddy the relations between near and far to create an indistinct zone or area in which particular entities are presumed to be present. Within these kinds of phase spaces, there are no clear boundaries that demarcate one area from another, which generates a sense of openness but also exposure for the entities that find themselves within that phase space. The Sun weather app is a good example of this diffused logic of modulation in action.

Sun is a weather app for iOS and Android smartphones. It uses the Dark Sky Application Programming Interface (API) to present weather data about a particular city or area that is selected by the user. Sun generates a diffused spatial intelligibility through the way that the app collects and presents data to the user. The Sun app is based on a core dashboard that displays the current weather conditions, both graphically and numerically. When opening the app you are greeted with a white circle that contains a temperature reading in digits. Above the white circle is a graphic icon that denotes the current weather conditions, such as cloud, sunshine or rain. A white line tracks outwards to the right of the circle and rises and falls, indicating changes in temperature across time. When the line is dotted, this indicates that rain is predicted and users can swipe to the right to move the circle along the line to check what the weather will be over the next twelve hours.

The Dark Sky API that powers Sun provides what it terms 'hyper local forecast and rain prediction with minute-precision at the "exact" location of the user' (Sun 2016: n.p.). The weather that is visualized on screen and the predicted forecast in Sun are therefore linked to where the user is located. Utilizing the Dark Sky API, the app is clearly designed to create a sense of empowerment for the user by providing them with the data to make supposedly informed decisions about the kinds of activity they undertake depending on the weather conditions in their area. But, at the same time, Sun also creates a phase space that discloses spatial relations for the user that are vague and diffuse because of the way this data is presented to and perturbs the user. For instance, when logging into the app from a smartphone, you can search for and select a city and the name of this city will then be displayed in the top portion of the screen. Sitting in Gosforth in Newcastle the app tells me that it is cloudy, but will rain in one hour. However, I have no idea exactly where the edge of this rain shower will be. If I travel to Benton will it be raining there as well? The way the app presents the data means that I cannot know unless I actually travel there and the app adjusts its forecast in relation to my position.

The logic of diffusion evident in the Sun app reflects broader trends in interface and app design, which aims to produce a sense of ease, elegance and simplicity. Commenting on the app, Rhodes (2016: n.p.) suggests that its 'design details make for an at-a-glance app that feels sensory and intuitive to use – looking at it feels a little like looking out a window, or stepping outside'. However, from a phase perspective, the app's design choices modulate intelligibility in which near and far become indistinct and unclear. On the one hand, the app provides very localized forecasting for the user, which they can use to make decisions about their activities. Seeing that it is about to rain in the next twenty minutes, a user may decide to stay in the house rather than take a walk for example. But, on the other hand, the Sun app also works to separate the user from the rest of the city or area where they are located. When using the app the rest of the city or area appears distant, because there is no way of making relations between near and far intelligible. The only point of spatial reference the user has is the name of the city at the top of the app and the rising and falling line, which does not denote spatial position. Without a point of reference or orientation, the user is left in the dark as to the diffusion of the weather system they are being exposed to, or might be exposed to in the future. In this case, the logic of diffusion is designed into the app's graphical aesthetic to create a simple and elegant experience, which is the app's key selling point. But, this aesthetic also works to modulate intelligibility in such a way as to cloud relations between near and far so that spatial relations become indistinct for the user.

b) Partition

The second logic of modulation is partition. Unlike diffusion, which aims to create a sense of spatial indistinctness, partition attempts to construct a sense of near and far as organized around a set of clearly delineated boundaries. The Nest Cam Alert system is an excellent example of the logic of partition in action. Nest Cam is a home security camera whose feed can be viewed live on a smartphone. As well as viewing the feed live, footage is saved on Nest's cloud servers for retrieval later. The camera uses a wide 130° angle lens, alongside a microphone and speaker to detect activity. Specifically, the camera 'determines when something is moving, based on how much an image changes over a short amount of time' (Nest 2017b: n.p.). In a similar manner, the microphone detects large jumps in noise against a background of ambient sound, which triggers an alert. Alerts are sent to a user's smartphone and appear as a notification. The user can then check the camera and even respond to activity using their voice, which is played back through the speaker on the Nest Cam. The logic of partition is utilized by Nest to emphasize the accuracy of the device and its ability to provide a simple way to protect your home from intruders. Video and testimony on the Nest (2017b: n.p.) website show clips of customers' homes being broken into, with the headline: 'Nest Cam helps solve crimes. Here's the proof', with customer testimony including 'I saw the intruders with the Nest app and used the Talk and listen feature to scare them away. Authorities were able to find and arrest the suspects thanks to the clip' (Nest 2017b: n.p.).

On a basic level, the Nest Cam discloses relations of near and far based upon the boundaries of the camera's field of view and where the camera is located. Nearness becomes whatever the camera's lens or microphone can be perturbed by and farness becomes what the camera or microphone cannot be perturbed by. But, this logic becomes more interesting in relation to the paid subscription service for the Nest Cam, Nest Aware. While the standard Nest Cam service saves images from alerts for up to three hours, Nest Aware records constantly twenty-four hours a day and saves this footage for either ten or thirty days. Furthermore, Nest Aware enables users to highlight what it terms activity zones. In Nest's (2017a: n.p.) words, 'with Activity Zones, you can define certain areas of interest in your camera's view and be notified when there's motion there, so the alerts become more … meaningful'. Specifically, the user can assign and monitor up to four different zones per camera, give each zone a custom shape, colour, size and name, receive specific alerts when activity happens in a zone and set up a zone 'where there's activity that you don't want to be bothered about, such as a window overlooking a busy view, but receive alerts for activity throughout the rest of the room' (Nest 2017a: n.p.).

Practically, this means the user can access a web interface and manipulate a rectangle with eight moveable points and attach this to a particular object in the frame of the camera. For example, the user may manipulate the rectangle so that it is the same size as a door in the camera frame. They can then add a name to this object, such as 'front door' and set the colour for this object (such as red, orange, purple, etc.). Through this process of manipulating a rectangle into a distinct shape and colour, each object becomes a partitioned zone, through which spatial relations between near and far are generated. As the user adds more zones, to doors, windows and other objects, multiple phase spaces become constructed. Each of these spaces only appear to the user of the Nest Cam when the camera senses movement within them and then alerts the user via a notification on their smartphone. Users can add different sounds for different zones, meaning that they do not even have to look at their phone to know when someone or something has entered that specific zone.

The logic of partition that underlies the phase spaces the Nest Cam and app construct is therefore not a matter of objects being partitioned or demarcated within a single or total space. Here, the Nest Cam could not be said to 'carve out images; reduce heterogeneous objects to a homogeneous space; and stitch together qualitatively different things such that attributes can be rendered quantifiable' (Amoore and Piotukh 2015: 344). Rather, the logic of partition constructs filters through which multiple phase spaces overlap and co-exist. For instance, users can set alerts for motion that occurs outside of the activity zones that they construct. Someone might walk past a Nest Cam without perturbing any of the activity zones and therefore not trigger an alert. Alternatively, you might set the coffee table and door as a zone, but then disable smartphone notifications for activity in the coffee table zone, while enabling notifications for activity in the door zone.

Through simple actions such as setting activity zones, the particular phase spaces constituted by the Nest Cam and app, which are organized around a logic of partition, modulate relations of nearness and farness for the user and potentially change the intelligibility of their homes. Rather than nearness and farness being organized around the objects that make up the house, such as sofas, curtains, pictures and so on, near and far become organized around the particular activity zones that a user applies in order to differentiate between objects and the alert notifications that are specific to each zone. Instead of nearness and farness being organized by the user's body as they reach across the table to pick something up or put something down, near and far become organized by the intentional and protentional structure of the Nest Cam, whose sensor is perturbed by degrees of movement. While operating on a micro level, generating differences in intelligibility are part of Nest's profit strategy. Nest wants people to become accustomed to using the activity zone feature, so they continue to pay for the Nest Aware subscription package. While sold as a safety feature that provides increased control for users, modulating spatial intelligibility into zones might also effect how home owners think about and experience space in less positive ways. For instance, using the app every day, perhaps the user begins to see their home as a space of potential burglary, with objects in the home understood through the way the camera is perturbed by movement that could potentially trigger an alert, rather than as objects offering a sense of comfort, security or family memories. In any case, the phase space of the Nest Cam and app clearly disclose spatial relations that can generate different intelligibilities and responses than when an individual does not use the Nest Cam.

c) Envelopment

Envelopment is perhaps the most fluid and dynamic logic of modulation that smart objects attempt to enable. Envelopment refers to the ability of some smart objects to disclose relations of near and far that dynamically respond to the movement and specificity of a user's body in order to create a sense of space and entities within that space as relative to an individual (on envelopment see Ash 2015a). Lumo Run is a good example of this logic at work. Lumo Run is a nine-axis sensor that clips to the back waistband of running shorts. The Lumo Run measures the runner's cadence (in terms of steps per minute), braking, (understood as changes in forward velocity), bounce, (understood as vertical oscillation), pelvic rotation, (meaning the rotation of the pelvis) and pelvic drop (referring to side-to-side drop of the pelvis) depending on the direction and force with which its sensor is perturbed. Using these perturbations the Lumo Run, in concert with a smartphone app, offers real-time feedback to users as they move to help them develop an ideal running form.

When using Lumo Run, it is possible to imagine that space appears as a kind of stretching and contracting envelopment. This envelopment emerges as relations of near and far shift around the target goals for each aspect of movement that the sensor is perturbed by. The ideal running form of an average adult is based upon moving in such a way as to ensure the sensor is perturbed as minimally as possible. For example, the target goal for bounce is less than three inches of vertical oscillation of the pelvis, the target goal for braking is less than 1.65 feet per second of slow down between steps, the target goal for pelvic rotation is less than fifteen degrees of side-to-side movement and the target goal for pelvic drop is less than twelve degrees of lateral movement across the pelvis (Lumo Run 2017).

Given these metrics, it would be easy to suggest that the spatial intelligibility of the Lumo Run user is organized around a process of quantification in which the 'body's activity … [is not understood] … in terms of the aching of one's legs or the amount of sweat issuing from one's pores, but in the numbers collected by pedometers, accelerometers, gyroscopes, and geolocative devices' (Gilmore 2016: 2534). From this position, spatial intelligibility could be said to emerge from the metrical recording of the movement of the user's own limbs, such as the length of their stride or the number of steps taken per minute, with the app encouraging the user to become increasingly aware of both the metrical height of each step and the metrical distance between each step through units such as centimetres.

But as Valentina Palladino, a user of Lumo Run reports, her spatial intelligibility was actively modulated through sonic notifications played via the Lumo Run smartphone app in ways that were not metrical or quantitative. In her words:

> The app gave me a goal of 158 steps per minute, and that wasn't as easy to achieve as it sounded. During my next recorded run, sound effects pumped into my ears – a high, congratulatory tone when my cadence was on-point and a low, sad trombone-like noise when it was lagging. By the end of that run, I never wanted to hear that low tone again. Because I wasn't used to running with such quick cadence, I heard the trombone nearly every two minutes. I also heard it incessantly because I chose the 'immediate feedback' option in the app – if the sensor knows you're not running properly to reach your goal, it will let you know with that sad tone immediately, and do so every time you fall off the wagon. Even though the sad sound effects made

> that run annoying, the feedback worked. Each time I ran after that, I found myself moving my feet more quickly to get more steps in per minute. When I recorded a run again with the Lumo Run, I heard very few sad trombones and many more happy chimes. It felt good. I'm not sure if my improvement was because I subconsciously didn't want to hear those negative tones, or I just got accustomed to running with higher cadence, but the method worked. Even after I stopped using Lumo Run, I still run with better cadence than I did before ... most of the time. (Palladino 2016: n.p.)

In this example, the way the Lumo Run is perturbed by the user's body discloses qualities of movement that would normally be imperceptible to the user as sound, which in turn changes how they make sense of the environment around them. While the logic of envelopment that Lumo Run employs promises the runner greater speed and efficiency and helps to avoid injuries associated with 'poor' form, this logic is less a matter of reducing qualitative bodily states to quantitative numbers and more a matter of altering the body's experience of spatial relations. For Valentina, relations of near and far became organized around the various qualities of sounds that she heard through her headphones in relation to her feet. Hearing happy sounds, she knew the spatial relations between her feet were 'correct', being neither too 'far' apart nor too 'close' together, while hearing the trombone, she knew she had to adjust the spatial relations between her feet, by increasing her pace or shortening and lengthening her stride.

Of course, just because the way Lumo Run attempts to modulate spatial intelligibility is not experienced as purely quantitative, does not mean that it is positive. For instance, relying upon the audio feedback from Lumo Run, relations of near and far are perhaps less organized around the particular objects on a run in a park such as the benches, fountains and play equipment that a runner might be running away from or towards and more organized around the individualized body and goal of the runner. In doing so, rather than enjoying the environment they move through, space potentially appears for the runner as set of bodily movements to be perfected and made more efficient. Like Nest and their Nest Cam, it is in Lumo's commercial interest for users to become habituated to the spatial intelligibility their Run device discloses because Lumo can then sell the user associated tracking devices and premium subscription services that offer them more ways of recording the perturbations associated with their run. But, unlike either Nest Cam or the Sun weather app, which create phase spaces where spatial relations appear as either partitioned or diffuse outside of the user, the Lumo Run modulates space to appear as individualized around the bodily movement of the users themselves.

The multiple logics of modulation

Smart objects generate phase spaces. These phase spaces in turn modulate the appearance of spatial relations in terms of nearness and farness, through which something like space appears to the humans and non-humans that find themselves alongside these phase spaces. As I have demonstrated with the examples of Sun, Nest Cam and Lumo Run, these phase spaces do not determine what people do. However, they do organize the disclosure of spatial relations around a series of distinct logics, which can influence how space becomes intelligible for users. These logics are not accidental, but actively designed to encourage users to become reliant upon these objects in their everyday lives.

The three logics of modulation identified in this chapter are not limited to Sun, Nest Cam and Lumo Run alone. The logic of diffusion can be identified in a range of smart objects, like smart vehicles such as the Google car, that intentionally perturbs objects using a range of sensors and cameras in order to create an intelligibility of space that appears to be wide, open and indeterminate.

This logic is the key to how Google car senses its environment, but is also used to market smart vehicles as having superior sensing capacities to human beings (see Chapter 7). The logic of partition can be further evidenced in objects such as the Nest smart thermostat, which allows the control of temperature on a room-by-room level. Here space is made intelligible as a set of distinct zones, which is used by Nest in their marketing materials to emphasize the personalization that their thermostat offers. Finally, the logic of envelopment can be recognized in apps such as Uber, which organize space and time around the customer using the app (see Chapter 6). This envelopment is designed to create a sense of spatial and temporal 'nearness' for the user in order to position Uber as the quickest and easiest ride sharing service to use and thus create and retain loyal customers.

While I have discussed the logics of diffusion, partition and envelopment in some detail, this is not to say that all smart objects and the phase spaces they generate modulate spatial intelligibility around these three logics alone. If we were to focus on a range of different smart objects and how they perturb and are perturbed by other things, undoubtedly many other logics could be identified. With this point in mind, the three logics discussed in the chapter should be considered as a starting point and orientating guide for further investigation of the spatial modulation of intelligibility rather than a definitive or complete list of logics that can be applied to all smart objects.

This chapter has argued that phases are primarily spatial constructions. But, this is only half the story. Chapter 4 suggests that phases of smart objects also make temporality intelligible. Phases disclose spatial relations of near and far, but also have a temporal durability that modulates how time appears as well. When phases are analysed as both spatial and temporal constructions, we can move on to investigate the political implications of phases (Chapter 5) and how phases can be actively altered or changed as a form of activism, through processes of what I term involution (Chapter 6).

4

Times

Work in the social sciences and humanities is increasingly pointing to changes brought about by the rise of smart objects in relation to memory, remembering and retention in everyday life (e.g. see: Greenfield 2006; Dodge and Kitchin 2007; Beer 2009; Kuniavsky 2010; Beer and Burrows 2013). For example, Kinsley (2015: 169) suggests that systems such as databases, social networks and Global Positioning Systems (GPS)

> demonstrate a potentially sizeable shift in the ways in which we govern, negotiate and understand our collective life. On the one hand, they place the potential for an extraordinary level of control over what is remembered, how it is remembered and what influence this can have on contemporary socio-spatial experience – where we can go, with whom we communicate and so on. On the other hand, these 'mnemotechnologies' are beginning to open out not only the access to but also the creation of shared, collective knowledge across a diverse spectrum of lives.

Beyond individual behaviour, within contemporary Western society 'cultural memory is stored in computational technologies such as online photo storage, document storage, etc., and also through the digitalization of culture with large-scale digital repositories of knowledge. In consequence, we are seeing a realignment of our contemporary culture … where new technologies … [are] … a site of materialized memory' (Berry 2014: 22), which become 'easily reproducible and distributed packages' (Parrika 2008: 71) that are brought, shared and sold.

Alongside a focus on the growing exteriorization of human memory into smart objects, there is an equal interest in how the same processes feed into and alter the structure of human memory as well. For instance, Stiegler suggests that contemporary smart objects, such as mobile phones and GPS act as forms of hypomnesis (Stiegler 2007, 2010b,c). By hypomnesis, Stiegler means that these devices stand in for human memory, to the extent that humans no longer have to remember key forms of information because the object itself remembers for them. For Stiegler this is largely a negative phenomenon because it creates passive and uncritical subjects. Simple examples of hypomnesis include systems such as GPS, which analyse and present a route to a user in real time so that they don't have to internalize or remember where they are going or how they will get there. In a similar vein Hansen (2015) and Raunig (2016) point to technologies that operate at the micro-temporal level in order to shape how users anticipate the future (also see Ash 2015a especially Chapters 4 and 7). Social media such as Facebook for instance works to anticipate what content the user wants to see based on their past actions, but also creates embodied and affective forms of anticipation in the user through alerts and updates presented as a timeline of events (Karppi 2015).

In examining these smart objects Hansen recognizes that technical time regularly exceeds human sensory capacities. In Hansen's (2009: 305) words, 'Time … is in no way tightly bound to human time-consciousness, but is rooted in a far more minimal and fine-grained structure of repetition that is, at some basic level, indifferent to the various temporalizations which mediate it for experiential regimes of all sorts.' Advancing Hansen's point, we can argue that smart objects are shaping temporality itself in ways that can't be reduced to the experience or perception of human beings. Developing a notion of phase time suggests that smart objects modulate the temporal intelligibility of both the human and non-human objects that perturb and are perturbed by smart objects. Phase time can be understood as the qualities disclosed by perturbations that create distinctions between past and present which, in turn, modulates how the future appears within a situation. Put in another way, the concept of phase time allows us to understand how smart objects can shape human environments and activities without necessarily being sensed, perceived or experienced by the humans whose activities are being shaped.

Smart objects shape temporal intelligibility by modulating the relationship between intention and protention, based upon what I term the phase memory of objects. For now, phase memory can be defined as the construction of durable perturbations between objects. In turn, these perturbations disclose qualities of objects that create the appearance of change. Examining the intentional and protentional structure of a variety of smart objects I demonstrate how they generate phase times and how this in turn shapes the temporal intelligibility of both humans and non-humans.

To make this argument, the rest of the chapter forms three sections. The section 'Phase Time' unpacks the concept of phase time and shows how time emerges from the perturbations between smart objects and introduces the concept of phase memory. The section 'Gradation, Dispersion and Dilation' demonstrates how the phase memory and intention and protention of smart objects shape the temporality that emerges in a phase by disclosing relations of then and now, past and present. To do this, the section explores the intentional and protentional structure of the Driftlight smart bulb, Dyson 360 Eye robot vacuum cleaner and the HeartWatch app for the Apple Watch. Finally, the section 'Spatio-Temporal phases' reflects on how the concepts of phase space and phase time can be brought together in order to analyse a range of different smart objects.

Phase time

Current approaches to technical time are often based on an inscriptive or retentional model of temporality (Sturken 2008; Garde-Hansen 2011; Özkul and Humphreys 2015; Saker and Evans 2016). An inscriptive or retentional model suggests that technical objects are seen to denote temporality via the way change is registered through the display of the marks and scars of time. For instance, take the work of Kirschenbaum (2008: 23) who suggests that 'deleted file information is like a fossil: a skeleton may be missing a bone here or there, but the fossil does not change until it is destroyed'. Or, as Chun (2011: 97) puts it, via processes of inscription, 'memory hardens information – turning it from a measure of possibility into a "thing" while also erasing the difference between instruction and data'.

Media theorists such as Stiegler (1993, 1998, 2007, 2010b,c) are an exemplar of a retentional perspective. For Stiegler, technical temporality is regularly understood as expressed and evidenced through inscription. In Stiegler's (1998: 224) words, technical objects are an 'apparatus of inscription' or the 'material inscription of … memory … mechanisms' (Stiegler 2010d: 4). For Stiegler, time comes to be measured, recorded and experienced through the way it is retained in or on technical objects as a kind of hardening, calcifying or fossilizing. A photograph or film cell retains the punctum of its exposure, a computer hard drive only works because of a past set of marks on its surface and so on.

Here, time is theorized as something that happens to a technical object, with the technical object being situated in a temporal frame that shapes it. For Stiegler, this broader temporal frame is based upon a geometric conception of space and an entropic model of temporality. In Stiegler's (2012: 32) words, retention is a 'spatialisation of time' predicated on three modes of co-existing temporality: physical, negentropic and technical. As he puts it,

> There is physical temporality, in other words, what we call entropy, the degradation of matter, the expansion of the universe – personally, I never call this temporality, I call this becoming. Then there is the temporality of the living, which as you know is a negentropic temporality, that is to say a temporality that fights against disorder … a structure that attempts to differ from and to defer … entropy, that fights against entropy. This is life. And there is a third temporality, the one that people generally call human temporality. But I don't like to call it human, because I think that it is an anthropocentric determination. I call it technics. It is a temporality within which a living being, in particular the one that we call man, is constituted in relation to the temporality of a technics which is itself a technical development or becoming. (Stiegler 2003: 155–6)

For Stiegler, temporality is essentially a process that is outside of an object, which the object has to fight against to continue to exist. Retentions or inscriptions are finite because any inscription has a limited temporal existence. A rock may be marked by a human tool, but left outside in the elements this mark will eventually be weathered away. A cherished subjective memory may be regularly reactivated within the brain of a human being, but this subjective memory is lost upon the death of the human. A computer hard drive fails and so on. Any inscription or retention is then subject to an ongoing 'retentional finitude' (Stiegler 2009: 188) that determines the limits of a particular retention's duration of existence.

While emphasizing physical time as a form of entropic becoming that is distinct from human or technical time, Stiegler's account of entropy still underlies his notion of technical and human time. It is the entropic nature of matter that forms the basis of the retentional finitude of objects through which humans come to form a relationship with time via the way technical objects are inscribed. Put in another way, time is not in the technical objects themselves, but is the entropic frame that all objects are ultimately governed by. With this point in mind, we could state that underlying Stiegler's retentional account of time is a kairological notion of change. In Boer's words, a kairological approach attempts 'to show the importance of the time of the event for the possibility of thinking change' (Boer 2013: 2). Steigler's account of time is kairological because any change in a body or object is understood in relation to the finitude of its retention, which is only finite because it degrades or breaks down via various events that are governed by the rules of entropy. In other words, change is ultimately enabled by a physical negentropic time as a temporal process that occurs exterior to a technical object.

Distinct from a retentional approach to technical time, phase time can be defined as relations of then and now that are made manifest by the qualities disclosed by the way smart objects perturb and are perturbed by other objects, which generate the appearance of change and difference. Here, time is not something that happens to a smart object, nor is time registered through smart objects via the way they inscribe or are inscribed by other objects. Rather, time emerges from smart objects themselves. A retentional account would suggest that temporal events or the passing of time brings about changes in objects. Instead, in a phase time account, change emerges from perturbations and qualities disclosed by an object or set of objects themselves, which creates the appearance of events as temporal. A phase model of temporal change does not assume that all the possibilities of what a smart object can do are present in an object's current state or emerge through a future event. Depending on how smart objects perturb and are perturbed by one another, smart objects can always disclose previously undisclosed qualities of an object, which shape what the smart object can do. Practically this means there is always an unknown potential for change that has its genesis in the interior of the smart object, or the object that is perturbing or being perturbed by the smart object itself, rather than a temporal system or force exterior to these objects.

From a phase perspective time itself does not 'move'. Instead, smart objects perturb other smart and non-smart objects, which creates the appearance of time. Linear, temporal succession, or the assumption that time moves in a straight line from past to present to future, is thus the outcome of the way objects perturb and disclose qualities of one another. Rather than a single linear time, smart objects disclose multiple, localized presents, pasts and futures. As Durham Peters (2015: 367) puts it, 'There is no single now that pervades the universe. Every now has a radius of dissipation, a broadcast footprint like a satellite. Now only stretches as far as our signals carry … no place in the universe is truly simultaneous with any other.' In turn, from a phase perspective, smart objects have what could be termed a form of phase memory. Instead of retaining inscriptions, as in the work of Stiegler and other media theorists, phase memory can be understood as the durability of qualities disclosed by the perturbations that are shaped by the intentional and protentional structure of smart objects. To unpack this idea in more detail we can productively turn to Ernst's understanding of computational memory. As Ernst (2013: 100) suggests,

> Memory is technically defined as a device into which information can be introduced and then extracted at a considerably later time. This kind of definition of time is close to what is known as a buffer in electronics. … It turns out that storage is nothing but a limit value of transfer. … Transfer and storage are two sides of one coin: storage is a transfer across a … distance. The traditional separation between transmission and storage become obsolete.

Non-human or technical memory is phase memory in the sense that all memory is a buffer – a transfer between two points – a limit transfer of value in Ernst's words. From a phase perspective, transmission and storage become 'two sides of one coin'. The memory of smart objects is not an inscription, holding or keeping, but rather the repetition of perturbations that disclose qualities of objects as more or less fixed or changeable. These repetitions occur within particular components of objects, between components of objects and between smart objects and other things.

To demonstrate this model of phase time and memory in action, think of the Amazon Echo multimedia speaker with voice control. The Echo uses Amazon's Alexa voice service to answer questions and fulfil tasks. Similar to the functionality of Siri discussed in Chapter 2, users can ask Alexa to play music, buy products from Amazon or check the weather by simply calling out the name Alexa while they are in range of the speaker's microphone. The Echo can also be used to control and receive updates from other smart objects, such as Fitbit fitness trackers, Philips Hue smart light bulbs and the Nest smart thermostat.

From a phase perspective, the Amazon Echo does not simply exist 'in' time, but creates a sense that time is passing through the way it perturbs and is perturbed by other objects that are shaped by its intentional and protentional structure. A simple example of this would be asking Alexa to add an item to the schedule on a digital calendar. Like Siri, Alexa is both intentional and protentional in the sense that its microphone can selectively differentiate between human speech and other sounds in the environment and it is constantly awaiting the word 'Alexa' to be spoken. In saying Alexa, the sound waves from the voice of the user perturb the microphone in the Echo, which then perturb the voice processor. The Echo then registers that it is processing the user's request by disclosing different qualities of colour on the top of its rim. Through this process, the Echo shapes the appearance of time, creating both a sense of temporal succession and that time passes at varying speeds.

How time is disclosed while using Alexa is also shaped by the various phase memories of the Amazon Echo. These are the repetitions of perturbations that disclose qualities of the components that make up the Alexa. If a perturbation continues to disclose the same quality of a component, this can generate the appearance of fixity or memory. If a perturbation discloses different qualities of a component of the Echo this generates the appearance of change. Depending on the number and rate of perturbations and the qualities they disclose, these changes appear faster or slower and this gives a sense that time is moving faster or slower. Some of the memories of these components are highly durable, such as the Texas Instruments TPS53312 Step-Down Regulator with Integrated Switcher. This component is perturbed by power from the electrical plug and discloses qualities of this power as the correct voltage that the Amazon Echo uses to operate. As long as the Echo is plugged in and not physically destroyed, this highly concretized component will continue to reliably repeat the same perturbations and thus create a sense of temporal fixity and stability as long as the Alexa remains homeostatic as an object.

The phase memory of other components of the Echo is less durable. For example, the firmware that provides the latest version of the software might repeat and disclose the same qualities of other components for one month or one week, until the next version is available to download. At this point the qualities this component discloses change as the object itself changes, which generates a sense of time as a distinction between the past of the previous firmware and the present of the current firmware. Finally, some of the repetitions of perturbation are very fleeting, creating a constant appearance of change. For instance, the lights that respond to users' voice requests consist of four Texas Instruments LP55231 Programmable 9-Output Light Emitting Diode (LED) Drivers. When the Echo's rim glows cyan, this means the microphone is being perturbed by sounds waves from a user's voice. When the Echo is failing to be perturbed by a router and is offline an orange light will circulate around the rim. When perturbing the router and cloud servers in order to analyse and process voice requests, the light cycles between two different shades of blue. Through these cycles of perturbation, the qualities of coloured light disclosed by the Echo create an experience of succession, in the sense that it generates the appearance that each perturbation and the quality it discloses are temporally linked to one another. At the same time, if

Alexa struggles to be perturbed by the voice of the user in such a way as to correctly analyse the voice and provide an appropriate response, this can also alter the speed at which time seems to pass. Being perturbed by the LED rim and waiting for a response, the user might become frustrated when asking Alexa to fulfil a simple task. Or conversely, if the Amazon Echo is connected to a strong Wi-Fi signal and the voice analysis servers are not being overly perturbed by digital sound files, then the blue light might only cycle once or twice, disclosing qualities that perturb the user in a minimal way. In doing so, these perturbations create the appearance that time passes quickly as the request is filled without pause or break from Alexa.

To summarize, the phase time of smart objects is the outcome of two interlinked processes. First, the appearance of qualities changing is shaped by the internal components of a smart object that are governed by its homeostatic and allopoietic structure, which affords it particular intentionalities and protentionalities. The qualities that create the appearance of change can be intentional in the sense that they are designed into the intentionality and protentionality of the object and contingent in the sense that the designer of the object can never fully know what quality of the object will be disclosed. Second, the appearance of change is produced by the ways smart objects perturb and disclose qualities with other things. Depending on how a smart object perturbs something else, qualities may emerge that also create an appearance of time as an apprehension of difference or change.

Just as the phase spaces generated by smart objects modulate the spatial intelligibility of humans and non-humans by disclosing relations of near and far, the phase times of smart objects also modulate the temporal intelligibility of human and non-humans by disclosing relations of then and now, past and present, which also shapes how the future appears. In addition to the logics of diffusion, partition and envelopment that were identified in Chapter 3 in relation to phase space, we can identify three logics of temporal modulation specific to smart objects. These can be termed gradation, dispersion and dilation. To understand how phase times operate to modulate the temporal intelligibility of humans and non-humans and discuss these logics in detail, we can turn to the example of three smart objects: the Drift light smart bulb, the Dyson 360 Eye robot vacuum cleaner and the HeartWatch app for the Apple Watch.

Gradation, dispersion and dilation

a) Gradation

A key logic of modulation in relation to temporal intelligibility is gradation. Gradation should not be confused with a logic of quantification, in which time is cut up into distinct equal parts expressed in technologies such as the clock, which measure time metrically (Bastian 2016). For example, writers such as Moore and Robinson (2016) suggest that smart objects can be understood as exemplars of quantification, through the way they enable second-to-second tracking of movement, which enables a neo-Taylorist model of selfhood. Distinct from the logic of quantification, gradation refers to increases or decreases of degrees of change, moving from a state of greater intensity to lesser intensity or vice versa. While quantification works to count and measure time as divisible, the logic of gradation is designed around the attempt to make time appear to move imperceptibly from one moment to another. The logic of gradation is exemplified in a smart object like Drift.

Drift is a smart light bulb used to aid the process of falling asleep. The manufacturer Saffron markets the product as providing a solution to changing lifestyles brought about by modern technology. As Saffron puts it, 'Before electric lighting, people's lives revolved around the natural light cycle of the sun. The rise of electric lighting and electronics means we don't have to depend on the sun for light, but this new lifestyle also means our bodies aren't able to prepare for sleep like they used to' (Saffron 2017b: n.p.). When activated, the

Drift uses LEDs in concert with a microprocessor to slowly dim over the course of thirty-seven minutes. Thirty-seven minutes is the time it takes for the bulb to complete its dimming process, because, according to Saffron (2017a), 'Any immediate shifts from bright light to darkness can … make it harder to get to sleep … [and] … 37 minutes is the average duration of a sunset. It gives your body the gradual shift from light to darkness that your body needs in order to prepare for sleep.' The Drift has two modes of use. In the midnight mode the light turns off after thirty-seven minutes. In the moonlight mode, the bulb dims over the same period, but remains on at a low level. Saffron claims that the warm yellow light given off by the LEDs avoids 'the bluer light emitted from electronics and some energy efficient bulbs … [which is] … especially bad for sleep' (Saffron 2017b: n.p.). Beyond Saffron's marketing claims, we can also argue that the Drift produces phase times that modulate the temporal intelligibility of the humans and objects that exist alongside the Drift.

The Drift generates phase times through the way it perturbs and is perturbed by other objects to create the appearance of temporal succession as gradation. These perturbations involve the LEDs inside the bulb, the microelectronic switch that controls the Drift's light output and the electricity that is used to power these elements. Combined, the LEDs in a single Drift give off five hundred and thirty lumens of warm light. As light is emitted from the Drift it perturbs objects that surround it, such as furniture, carpet and curtains. In doing so, various qualities of these objects are disclosed, creating shadows, reflections and altering the appearance of their colour. The warm light from the LEDs might make white objects such as a sofa look yellow or a purple object such as a vase look brown or black. As the microprocessor perturbs the LEDs, disclosing dimmer light, this creates the appearance that time is passing. This is confirmed as the appearances of objects continue to change. Shadows grow longer and for a human in the room, objects become harder to identify and differentiate between as the bulb discloses fewer qualities of the objects it perturbs.

Based on these perturbations, it would be possible to state that the Drift discloses multiple phase times for the various objects that are perturbed by the light from the bulb. These times are organized around the intentionality of the microprocessor that measures thirty-seven minutes when the Drift is activated and the gradation at which the light level drops from a full five hundred and thirty lumens to either darkness, or a low light. In doing so, the temporal intelligibility of the room that the Drift discloses modulates the relationship between past, present and future for the various objects that find themselves in the room. For instance, a human lying in bed might be reading a book with the midnight mode activated and use the midnight mode to determine when to go to sleep. Rather than looking at a clock, the relationship between past, present and future is organized by the gradually fading light, which casts longer shadows in the room and makes their book harder to read. In doing so, the changing qualities of objects in the room that are disclosed by the perturbations of the bulb shape relations between past and present that modulate how the human lying in bed anticipates a coming future of bedtime and remembers the room being well lit when they began to read. As this example shows, the phase time that the Drift generates, alongside the associated relations between past and present that it makes intelligible, is not just in the Drift and how it perturbs things. The phase time is also in the qualities that are disclosed by objects that are perturbed by the Drift.

Beyond the conscious awareness of the human user, the phase time of the Drift can also modulate the temporal intelligibility of users in ways that do not enter into conscious awareness and are perhaps not even phenomenally experienced at all. A simple example of this would be the way the Drift can potentially alter the production of melatonin in users' bodies and shape their circadian rhythms. Melatonin is a hormone produced by humans and animals that helps regulate cycles of sleeping and wakefulness. Levels of melatonin in the body are influenced by the amount produced in the body's pineal gland as well as through exposure to environmental light (Masana and Dubocovich

2001). Blue light produced by many LEDs and smart objects such as phones and tablets suppress the production of melatonin, compared to other types of light, thus making it harder to rest or sleep (West et al. 2011; Gabel et al. 2013). The production of melatonin is one part of the body's broader circadian rhythms, which can be understood as the physical and mental changes that follow a twenty-four-hour cycle in response to light and darkness. The Drift utilizes yellow LEDs, which do not suppress the production of melatonin as much as blue LEDs, which are used in many other bulbs. This allows Saffron to claim that 'by reducing the amount of blue light emitted … [the Drift] … increases melatonin and helps get you ready for sleep'. Imagine an insomniac began to use the Drift and found the yellow light that gradually darkened enabled them to sleep regularly. Here it could be argued that the temporal relations between past, present and future disclosed by the Drift perturbed the body, disclosing different qualities of various chemicals that regulate the circadian rhythm of the user on a biological and neurological level. As such, the phase time of the bulb modulated the temporal intelligibility of the objects that were perturbed by it, regardless of whether these perturbations or the qualities they disclosed were consciously perceived.

The way the Drift modulates temporal intelligibility is key to the commercial logic of Drift as a product. While it would appear that Saffron's goal is admirable in helping to ensure better-quality sleep, at the same time, the logic of gradation could also potentially lead to issues. For example, Van den Bulck (2015) suggests that the rise of smart objects linked to activity, well-being and sleep can actually create new forms of health issue such as chronorexia. In Van den Bulck's (2015: 123) words, chronorexia refers to 'people who develop an unhealthy and, possibly damaging, obsession with "healthy sleeping," as measured by their devices, either in themselves or in their children'. With this point in mind, as with all the other logics of spatial and temporal modulation discussed across this book, it is important to remember that just because the phases smart objects generate do not necessarily quantify or abstract experience it does not mean they aren't designed with commercial logics in mind and cannot have problematic effects.

Regardless of the outcome, the phase times of the Drift are not a form of metrical time in terms of hours, minutes or seconds. As we have demonstrated, phase times are the qualities disclosed by the intentional and protentional structure of smart objects that shape how they perturb and are perturbed by other objects, which create the appearance of difference or change. From this position, time is not a process that changes objects, but is the very qualities that are disclosed by those objects as they perturb and are perturbed by one another.

b) Dispersion

Alongside a logic of gradation, the intentional and protentional structure of smart objects can also modulate time according to a logic of dispersion. While gradation is a matter of making the movement of time imperceptible, dispersion is an attempt to make time disappear completely, so that time is not intelligible at all. Dispersion involves attempting to minimize the qualities disclosed by smart objects as they perturb and are perturbed by other things. We can see this logic at work in the Dyson 360 Eye robotic vacuum cleaner. Released in 2016, the Dyson 360 Eye is an autonomous vacuum cleaner that moves around on tank tracks and uses sensors to automatically hoover up debris from floors, without direct human intervention.

On the surface, the phase time of the Dyson would seem to be organized around a single form of quantitative clock time. The smartphone app for the device allows you to set a schedule of cleaning based on days of the week and times of the day. When activating the 360 Eye using the app, time also appears to be spatialized through a map that visualizes the movement of the 360 Eye, which is presented alongside a record of how many square metres it has cleaned, the number of times it recharged while completing a cleaning task and its total running time, presented in hours and minutes. However, examining the varying qualities of multiple objects the Dyson discloses through its perturbations, it becomes clear the Dyson is not organized around clock time, but actively attempts to modulate temporal intelligibility around a logic of dispersion to make the time of the device disappear for the user.

The 360 Eye uses a 360° camera and infrared sensors to navigate around areas where it is deployed and sucks up dirt and debris using a rotating brush in concert with a cyclone motor. The 360 Eye contains two power modes, which allow it to operate between forty and seventy-five minutes at a time. When not in use the 360 Eye charges on a charging dock and automatically returns to the dock when it senses its battery is running low. The 360 Eye generates a phase time through the ways its sensors perturb and are perturbed by other objects. In turn, these perturbations create the appearance of temporal change or difference. The key perturbations of the 360 Eye are organized by its 360° camera, infrared sensor and Simultaneous Localization and Mapping (SLAM) software. The 360° camera produces a panoramic image, which allows the robot to 'see' all around itself without having to turn. This saves battery and ensures that it can't easily be 'blinded' (as it might do when bumping into something if it was using a forward-facing camera). Using 360° 'vision', the camera identifies at least three high-contrast objects within the image (such as the edge of a chair, a worktop and a picture frame) and uses these to triangulate its position in the room. The 360° camera means the same objects can be tracked in the image without ever going out of shot, ensuring that the robot does not lose its orientation.

In concert with the 360° camera, the Dyson also uses a series of infrared sensors. These sensors allow it to detect its distance from other objects and voids so it doesn't fall down stairs or drop off obstacles. Utilizing the qualities disclosed by the perturbations of these sensors, the 360 Eye draws upon SLAM software to control its navigation and movement. Once it has scanned a room, the 360 Eye will move to the centre of the room and then travel outwards in rectangles. The 360 Eye is programmed so that it travels one robot's width of where it has been before. This creates a square spiral pattern of movement. As it moves it both maps and records its movement, meaning that 'it knows where it has covered and it knows where is left to be cleaned' (Dyson 2017: n.p.). This map is visualized on the 360 Eye's smartphone app, which creates a solid blue colour for where it has cleaned, a dark blue colour that denotes walls and other boundaries and a white colour that shows the presence of an object such as a coffee table. Once the 360 Eye has mapped and cleaned all the available space, it then deletes the map and resets its memory so it can be used in a different environment the next time it is activated.

The Dyson 360 Eye's entire intentional and protentional structure is therefore designed to minimize human involvement and it attempts to achieve this partly by creating a sense of temporal diffusion. What is key to the logic of diffusion is how smart objects modulate the appearance of time, which is often more important than reducing the quantitative amount of time a smart object might take to complete a task or carry out an activity. For instance, robot vacuum cleaners do not necessarily save time compared to using a push-along vacuum cleaner. As Sung et al. (2007: 150) point out in their study of the Roomba robot vacuum cleaner, the Roomba

> did not save householders' time and labor because it both took time, and also created monitoring and maintenance tasks. Participants described that cleaning with a Roomba took longer than with a traditional vacuum cleaner – albeit in smaller chunks – because of the need to move the machine around the rooms of the house. … They also … told us about their Roombas getting stuck underneath chairs or trapped in the bathroom … [and] … described how brushes, bins and motors needed cleaning to remove the fine dust that might corrupt the sensors and affect Roomba's function.

In a similar manner, reviews of the 360 Eye have pointed to the considerable length of time it can take to clean the floor of a house, as the Dyson needs to regularly return to its base station to recharge (Bell 2016). As long as the user is not perturbed by the qualities the 360 Eye discloses as it goes about its business, then we might assume that it does not matter to Dyson if the 360 Eye actually saves a metrical 'amount' of time or completes its task more 'quickly' than a regular vacuum cleaner. In other words, the example of the 360 Eye demonstrates that modulating the appearance of time can be more important to the manufacturers of smart objects than how the device's activities are measured in relation to a series of quantitative or metrical temporal units.

c) Dilation

While the Dyson 360 Eye is designed to disperse perturbations to make time disappear or move to the background of intelligibility, other smart objects attempt to organize intelligibility around a logic of dilation. Dilation points to how smart objects can perturb objects in such a way to make time appear to move slowly, or bring time to the forefront of intelligibility. A simple example of the logic of dilation in action would be the HeartWatch app for the Apple Watch. The app uses the heart rate sensor on the back of the watch, alongside accelerometer and GPS data to track how the user wearing the watch moves and how their heart rate rises and falls as they move. According to Apple, the heart rate sensor uses the medical technique of photoplethysmography to measure a user's heartbeat. This technique detects the movement of blood through the wrists by flashing a series of coloured LED lights onto the skin. In Apple's (2017: n.p.) words,

> Apple Watch uses green LED lights paired with light-sensitive photodiodes. … When your heart beats, the blood flow in your wrist … and the green light absorption … is greater. Between beats, it's less. By flashing

> its LED lights hundreds of times per second, Apple Watch can calculate the number of times the heart beats each minute – your heart rate. In addition, the heart rate sensor is designed to compensate for low signal levels by increasing both LED brightness and sampling rate.

Here, the phase time that the watch discloses is organized around the number of times its heart rate sensor perturbs the user's wrist and is perturbed by the blood in the wrist. In doing so, the sensor discloses different qualities of colour of the user's blood as it is perturbed by white light from the LEDs in the sensor. Through this sampling technique, the user's heartbeat becomes measured in relation to metrical time as a Beats Per Minute (BPM) reading. The HeartWatch app displays this BPM reading in a number of ways and across four 'views' that are displayed visually on the app on the watch's face, which it terms waking, regular, workout and sleeping. Each view is kept separate so the app can measure your heart rate in relation to an appropriate average. For instance, if you are working out, you can select the workout mode, which would set the average heart rate higher and if seated you set the app to waking, which would set the average heart rate lower. One way the heart rate is displayed within these views is through a round badge, which consists of different coloured circles depending on your activity state. In the regular state, the badge consists of blue, purple and red circles, denoting resting, high resting and elevated heart rate states, respectively. This badge gives the user an overview of their heart rate over the course of a day. According to the makers of the app,

> the colours represent time in zones overlaid with an average. For regular heart rate readings, the ideal is to have a completely blue badge. The bigger the red ring around the badge, the more time spent with an elevated heart rate. Even though averages can appear to be similar, the badge shows a clear picture of how your heart has really been tracking and alerts you about excessively high or low readings. (Tantissia 2017: n.p.)

Below the badge is a reading showing the lowest and highest heart rates of the day and the exact time these readings were taken. Through the phase time generated between the heart rate sensor, app, watch and user, time becomes intelligible as a kind of dilation. Humans are usually unaware of their heart rate or heart beat at all, but the HeartWatch badges make this bodily function highly visible to the user. In turn time becomes organized around the user's heart rate in relation to a BPM calculation and metrical clock time, which splits the day into hours, minutes and seconds. Looking at the badge and associated data on the watch screen, each minute or moment becomes more pronounced and apparent. If the user also looks at the badge to identify the visual relationship between resting, high resting and elevated states and when these states were recorded, these moments might take on an increased significance for the user. For instance, the user may reflect on what they were doing at this time and try to identify the particular event that might have triggered an elevated heart rate.

This logic of dilation becomes more pronounced when users access live readings of their heart rate through the app. Here, the heart rate is displayed on a speedometer-style gauge composed of distinct sections of colour, from light pink to blue to purple to bright red with a white needle pointing to the current heart rate. When accessing the live reading by pressing the side button on the Apple Watch, users can see their current heart rate and quickly ascertain whether it is too high or low. Furthermore, the app can also alert the user if their heart rate is too high or low, through a text notification on the watch, even when the user is not actively monitoring their heart rate. Here, the phase time of the Apple Watch and app dilates time by sensitizing the user to small units that are organized around the perturbations of their own body in relation to the watch's heart rate sensor. In doing so, temporal intelligibility becomes modulated around potential health issues and bodily states of stress.

Lupton (2014: 612) suggests that health apps accessed on devices like the Apple Watch 'offer … opportunities … to engage in obsessive self-surveillance because of their capacity to produce detailed data continuously in real-time'. Distinct from Lupton's account, I have demonstrated that there is no real-time flow of data to obsess over. Rather, the Apple Watch and HeartWatch app modulates time itself by disclosing a greater or smaller number of qualities, so specific moments appear as larger or smaller to the user. How the Apple Watch and HeartWatch app shifts users' intelligibility to make time appear bigger or smaller can have a number of effects. Users may become more aware of their heart rate and more wary of engaging in activities that raise their heart rate. At the same time, the app might create a sense of anxiety as the heart badge denotes that the user has had an elevated heartbeat across the course of a day, which the user may never have been aware of without the app. The makers of HeartWatch employ the logic of dilation in order to provide information that they suggest can improve health, aid exercise and alert users to potential health problems. But, this same logic can also potentially create anxiety and encourage users to focus on micro-temporal changes in their heart rate in ways that are actually unhealthy and problematic.

Spatio-temporal phases

This chapter has focused on the Drift smart light bulb, Dyson 360 Eye vacuum cleaner and the Apple Watch and HeartWatch app in particular to make the argument that smart objects generate phase times, defined as the changing qualities of objects as they perturb one another. The intentional and protentional structure of smart objects produce phase times that attempt to modulate the temporal intelligibility of users by shaping the relationship between past, present and future. Through this analysis we identified three logics of temporal modulation: gradation, dispersion and dilation. In the examples discussed here, these logics were used for explicitly commercial ends in order to sell products. In the case of the Drift, its main selling point is that it can modulate the intelligibility of users by creating the appearance of temporal gradation to help them get to sleep. In the same way, the Dyson 360 Eye is sold as a labour- and time-saving device that modulates time to appear as dispersed and backgrounded from human awareness. Employing the opposite logic, the HeartWatch app on the Apple Watch is sold on the fact it can help people become aware of potential health problems, by modulating time to appear at the forefront of intelligibility.

While the examples of the Drift, Dyson 360 Eye and Apple Watch HeartWatch app are very specific, many other smart objects can be understood as generating phase times that modulate temporal intelligibility according to the logics of gradation, dispersion and dilation. In relation to gradation, we can think of objects such as the Hive smart heating thermostat. The Hive is designed to modulate temperature in accordance with outdoor temperature, user preference and a number of other factors so that the temperature of owners' homes alter without sudden shifts from hot to cold or cold to hot. In doing so, time becomes intelligible through gradual shifts rather than distinct cuts or jumps between perturbations. With regard to dispersion, think of the Smartpot, a smart plant pot. Smartpot waters plants and provides plant food automatically, dispersing temporal intelligibility to the point at which the user only has to anticipate re-filling the pot once a month, rather than provide daily or weekly watering manually. In terms of dilation, think of smartphone apps that provide 'real-time' updates of bus, metro and train times that are available for transport networks in cities around the world. For instance, the Nexus Metro app, designed to complement the metro train system in Newcastle, UK, allows users to input information about what stations they are travelling to and from and then counts down the time in minutes until the first, second and third trains due to arrive. Rather than relying on older forms of static timetables, using the app to plan journeys alters the temporal intelligibility of the user and the metro system. With no leaving times provided, movement can become organized around a heightened awareness of when the next train will arrive, which is divorced from forms of clock time that are related to fixed or printed timetables. Within the phase times of the app, the duration of a user's journey to a station can take on a new anticipatory intensity as they watch the minutes tick down before the next train arrives. In doing so, the phase times of the app discloses relations between past, present and future in which time is individualized in relation to the user's goal and the Metro trains as a commercial service.

While I have split the discussion of phase space and time into separate chapters for analytical purposes, it should be clear that phases are both spatial and temporal. If we take the logics described in the previous two chapters seriously, this results in an account of multiple spatial and temporal logics that modulate intelligibility. For instance, Lumo Run attempts to modulate spatial intelligibility around the logic of envelopment and temporal intelligibility around the logic of dilation, while Sun attempts to modulate spatial intelligibility around the logic of diffusion and temporal intelligibility around the logic of gradation. In the same way that diffusion, partition and envelopment are only three possible ways smart objects modulate space, gradation, dispersion and dilation are only three possible ways smart objects modulate time. Following the mechanological impulse that guides our analysis of smart objects, I would encourage readers to investigate other smart objects that the book has not examined in order to identify their sensing components and name other logics of spatial and temporal modulation. Even with the six logics that I have identified so far alone, this results in hundreds of possible combinations of spatio-temporal modulations that smart objects can attempt to enable.

In Chapter 2, I argued that smart objects are best understood as having an intentionality and protentionality that shape what they can do. In Chapters 3 and 4, I have demonstrated how the intentionality and protentionality of different smart objects disclose spaces and times that attempt to modulate the spatial and temporal intelligibility of humans who use or live alongside these objects according to a variety of logics. With these points in mind, Chapter 5 turns to the politics of phases. While smart object manufacturers attempt to create phases that modulate spatio-temporal intelligibility to encourage people to buy and use their products, I argue that the politics of phases is far more contingent and open than the logics presented here might suggest. Phases disclose new domains and forms of contestation organized around what I term endo and exo politics and these politics can have powerful effects on both humans and non-humans alike.

5

Politics

A number of writers have begun to think through the politics of smart objects and the networks and infrastructures of devices that support them (Amoore 2013; Kember and Zylinska 2015; Amoore and Piotukh 2016; Bratton 2016). These accounts tend to focus on how smart objects constrict the possibility for some form of collective human life to take place (Coley and Lockwood 2012; McCullough 2013). The problem with smart objects, we are told, is how they work to redistribute control and power, creating individualized and privatized subjects, which limits the potential for human community and democratic or equal forms of being-together. This argument is common in a variety of work that deals with 'cloud', 'elemental' and 'ambient' networks of smart objects. For example, McCullough suggests that smart objects, which he terms ambient media, generate a series of problems for collective human life. In particular he asks:

> Does having more ambient information make you notice the world more, or less? Can mediation help you tune in to where you are? Or does it just lower the resolution of life? Today ambient information media become more difficult to escape. They channel more kinds of communications into shared physical contexts that, like city parks, come with expectations of being commons. (McCullough 2013: 273)

The implication here is that by over-coding public spaces with information, the possibility for a public sphere is threatened. Deleuze's (1992) work on the society of control is often cited in these accounts to show how smartphones and other technologies are creating a situation where people do not need to be spatially or temporally contained in order to be subject to state or corporate power (e.g. Bratich 2006; Tuters and Varnelis 2006; Best 2010). For example, Kember and Zylinska (2015: 127) suggest that ambient technologies in the home, such as computational smart surfaces built into worktops 'operate … as a form of productive containment, closing down on the potentiality and temporality of subjects and retroactively transforming the potentially fluid and metamorphic self into the marketized self, the becoming data machine'. Developing a similar sentiment, Hayles (2009) suggests that RFID tags create new forms of 'distributed cognition' that enable increasingly detailed tracking of individuals. Or as Hu (2015: XV) argues in relation to cloud computing, 'Deleuze's description of data aggregation, the amorphous and open environment of computer code and even the gaseous qualities of corporations within a control society map directly onto attributes of the cloud.'

Despite this statement, Hu suggests that understanding cloud networks through the concept of control is perhaps too simple. 'If we look at the cloud closely, we find … [that] … the all-but-forgotten infrastructures that undergird the cloud's physical origins, for example often originated in a state's military apparatus … [and] … today's cloud relies on a repurposed version of this infrastructure' (Hu 2015: xvi). Hu (2015: xvi) suggests these traces 'are a clue that the supposedly anachronistic modes of sovereign power may be returning under different forms'. Instead of developing an account of control, Hu (2015: 146) suggests that data has a sovereignty 'which describes the variable ways that sovereign power interfaces with data-centric tools', deciding 'what counts as human, who is included, who is excluded, when the protection of the law can be withdrawn, and therefore who can be discarded or killed'. From this perspective, the technologies and data that drive smart objects act as a mediating point where 'sovereign, disciplinary and governmental power constitute' three sides of a shifting triangle (Hu 2015: xvi). Although Hu critiques writers who employ Deleuze's concept of the control society, he still reiterates McCullough's claim that smart objects generate individualized modes of human subjectivity and action and that this individualization is problematic. In Hu's (2015: 147) words, 'The cloud produces users rather than publics, and therefore individual rather than collective action.'

McCullough and Hu are not alone in questioning the political implications of living in environments filled with smart objects. For example in *Program Earth*, Gabrys argues that technological sensors in cities constitute a new kind of 'withness', a 'becoming together, of concrescing, such that the possibilities for both urban ontological engagements as well as urban speculative futures are undertaken' (Gabrys 2015: 242–3). This withness might involve sensors that provide smart lighting as someone walks along a darkened street, or fitting animals with RFID tags to track their migration patterns in order to aid their survival and reproduction. Despite the potential of being-with, Gabrys (2015: 261) points out that although smart objects 'might admit some contingencies in the form of local circumstance … [they] … move toward the same end point of managing and regulating cities in order to achieve efficiencies and solutions'.

What is common in these accounts of smart objects is that the production, recognition and protection of collectivity between humans and other humans or humans and non-humans is seen to be the desirable aim of political thinking and action. Whether explicitly stated or not, these forms of politics tend to suggest that smart objects produce individualized and alienated subjects. In turn, smart objects are critiqued as part of a broader process of neoliberalism as a fundamentally negative set of rationalities, logics and techniques. As Davies (2014) suggests, neoliberalism is the idea that individuals should take responsibility for their own actions according to the logical abstractions of a market model of exchange. Under neoliberalism life becomes a kind of competition, where individual actors are encouraged to concentrate their efforts on themselves, dissolving social and communal ties and exacerbating inequality.

For theorists such as Couldry (2012: 73), media in all its form is considered to amplify such processes, by normalizing 'values and mechanisms important to neoliberalism'. From social media to televised singing and talent contests, media producers position and encourage individuals to understand themselves as a kind of product that only has value in terms of monetary worth. In doing so, neoliberal market economics 'reinstall[s] and reinforce[s] a traditional, humanist, solipsistic, and antimachinic vision of the self. When the transformation of individual subjects does take place … it is mainly on the level of their repackaging as data: a process that remains clearly inscribed within a market rationality' (Kember and Zylinska 2015: 129). As Couldry, Hu and a range of other writers suggest, mobile and networked social media, which are accessed via smart objects such as smartphones, only exacerbate such amplification by creating the illusion that individuals have a voice and that this voice matters. Jodi Dean (2005) echoes this point by arguing that encouraging users to upload selfies to Instagram, or update their status on Facebook from their smartphones, creates fantasies of participation that actively work to foreclose political critique or action. For Dean, clicking an online petition or uploading a blog post creates the appearance of political intervention, but this intervention is always mediated and shaped by networks that bury such action within a stream of data, where its distinct potency for change is lost.

Different from these accounts, I want to argue that phase politics is not a matter of critiquing particular companies that attempt to generate or modulate spatial and temporal intelligibility to control behaviour or create profit. Rather phase politics refers to the processes through which entities appear in the world as potentially political and around which political responses can be organized.

Phase politics thus points to the contingent nature of smart objects and the phases they create, which always exceed corporate control and manipulation. In other words, an account of phases suggests a cautiously optimistic reading of the politics of smart objects. Smart objects and the phases they generate open the potential for new forms of politics to emerge, rather than simply regulating and modulating human behaviour within a set of exploitative state or corporate logics.

To make these claims the rest of the chapter forms four sections. The section 'Smart Politics' suggests that the logic of arguments around the politics of smart objects is structured through a distinction between what Harman (2014) calls up and down or power and truth politics. Rather than pursue a model of power and truth politics, the section 'Object Politics' develops Harman's reading of Latour to posit a politics of smart objects themselves. Drawing upon, while differentiating myself from, Harman and Latour, in the section 'Endo and Exo Politics' I show how the concept of phases shifts the terms of debate to a consideration of what I term the endo and exo politics of smart objects. The section 'Phase Politics' unpacks the concepts of endo and exo politics through the specific example of the controversy surrounding electromagnetic radiation used to transmit mobile signals in the village of Wishaw in the UK. In the concluding section I suggest that endo and exo politics offer different ways of thinking from the critical approaches that have dominated political debates around smart objects.

Smart politics

Writers who critique smart objects as an expression of neoliberal forms of power, such as McCullough, Hu, Gabrys and others could be positioned on the left as opposed to the right side of the Western political spectrum. However, they could also be positioned along another political spectrum that Harman (2014) terms the distinction between up and down politics. In *Reassembling the Political*, Harman offers a reading of Latour's ANT in relation to politics. He suggests that Latour's work allows us to recognize the existence of formats that structure traditional distinctions between the left and right. Harman suggests that left and right political positions are organized around a more basic form of what he terms up and down politics. Put simply, up, or power politics refers to the belief that political judgements or decisions emerge from a struggle between different systems of knowledge and values. Down, or truth politics focuses on the desire to base political decisions on truths that are valid and immutable. For example, a true politician might argue that all humans have insurmountable rights that cannot be broken for any reason. Distinct from this, a power politician may argue that human rights can be broken if the situation demands it and the legal system can accommodate such a breach. For the truth politician a detained terror suspect must be treated like any other prisoner, whereas for the power politician, the terror suspect could be mistreated in order to potentially get them to provide information that would prevent an attack that might kill innocent civilians.

Interestingly, Harman suggests that versions of up and down politics are not tied to particular left- or right-wing movements or groups. From this perspective, one can be left wing and believe in power politics. Alternatively you could be left wing and believe in truth politics and vice versa. For instance, you might hold the position of right-wing power politics in the sense that you believe that the struggle between humans is natural and that the strongest should be the ones who take power. Or you may express right-wing truth politics, in the sense that you recognize that decisions should be based on appeals to principles of truth, but these truths should be expressed and realized at all costs. Positioning Hu, McCullough, and Kember and Zylinska on Harman's political compass produces some interesting insights. For Hu, politics in relation to the cloud technologies that power the majority of smart objects is a matter of control over resources that disable the potential for commons to emerge. In Hu's (2015: 147) words,

> The cloud is actually a metaphor for private ownership. What gathering spaces emerge within the cloud are closest to the ambiguously named privately owned public open spaces found in American city's financial districts. Though the landscaping of these pocket parks and rooftop gardens bears a superficial resemblance to public space, these overly tidy zones are nevertheless administered by banks, insurance companies and the like. The eventual consequence is that the lived knowledge essential for imagining and discussing public space has begun to atrophy.

In the same way that urban public space has become privatized, Hu suggests that the cloud is a key technology for closing down the internet as a digital commons or public space. To respond to this closing down, Hu (2015: 148) suggests that the cloud needs to be returned to 'the scarcest space of all: the space of public life'. While Hu's genealogy of the cloud and account of power is complex, his political response is straightforward. Undoing the sovereign power of the cloud requires breaking up large corporations that control the flows of data within a system and returning the cloud to public ownership. On Harman's spectrum this is a clear example of left and down politics. As Harman (2014: 2) puts it, left and down politics

> consists in the revolutionary view that humans are equal as thinking things and as bearers of inalienable rights. If humans do not currently enjoy such rights … it is because they are blocked by some ulterior force. … Under this model the key political act is opposition since the existing state of power will almost never coincide with truth, and must therefore be confronted or replaced.

This left and down position is echoed in McCullough's response to the rise of ambient smart objects. For McCullough (2013: 273),

> The considered life requires a balance between the messages and things, between mediated and unmediated experience. Citizens have a right to engage one another and the built world they inhabit in ways that are unmediated, uninstructed, unscripted, and undocumented. Under an ethics of preserving and protecting existing information environments, there should be a right to preserve the subtle high definition of the intrinsic structure of the local world, to protect it from being covered over with the crude low resolution of one-size-fits-all media productions.

Much like Harman's account of down politics, this position is based on an understanding of 'citizens' having particular rights that need protecting through the preservation of environments that speak to and elide with basic human capacities for attention. At the same time, McCullough implicitly points towards an essential account of human nature that is most able to flourish when insulated from the filters of smart objects.

Distinct from Hu and McCullough's left and down politics, Kember and Zylinska (2015: 154) point towards a politics of smart media 'where the agency of ethical subjects and objects is yet to be determined and where there are no predefined fixed values to be applied or drawn upon'. In their words, this 'ethics of mediation will … be nonnormative, in the sense that it will not resort to predefined values or truth in advance … nor will it posit a moral core from which an ethical judgement can be issued' (Kember and Zylinska 2015: 154–5). In relation to Facebook they suggest that such a political approach is not to reject social networking sites, but to 'become-different from them' (Kember and Zylinska 2015: 172). To do this, they suggest using Facebook, but always being mindful to introduce 'cuts' that break up the flow of engagement that Facebook attempts to foster. In Kember and Zylinska's case, politics is left and up. Here politics, in Harman's (2014: 3) words, is a 'war of all against all in which seizing power for one's own standpoint becomes an end in itself. The left version of the up standpoint is familiar from the identity politics of postmodernist intellectuals and from claims that desire is infinitely creative and must be subject to no sublimating social constraint.' For Hu, politics is a matter of unblocking individuals' access to basic forms of communal being-together, whereas for Kember and Zylinksa, politics is a matter of recognizing that there is no originary human state to be fought for and thus humans must work alongside technology to live with smart objects in a different way.

Although I have only discussed four different political responses to smart objects and there are important differences between these responses, such positions reflect many other critiques of smart objects as well. For instance, work on quantified self devices regularly suggests that these objects are central to abstracting and quantifying life according to a set of reductive logics. Moore and Robinson (2015: 2786) summarize such a position, when stating that self-tracking smart objects 'undermine … life to capital to an unsustainable degree, destroying the qualitative outside, which … provides the basis for capitalism (as use-value, labour-power, consumer desire)' (for similar critiques of the quantified self, see Lupton 2015; Gilmore 2016; Pantzar and Ruckenstein 2015). These responses would fit into Harman's political compass of left and right and up and down, with the majority fitting into the left and up or down category. Indeed, regardless of whether these thinkers are left and up or left and down, smart politics is more or less a matter of identifying the processes, techniques and procedures that technology companies use to create or exacerbate existing neoliberal conditions. Even Kimber and Zylinska (2015: 172), who posit a non-normative account of smart objects still argue that smart politics is ultimately about the 'decoupling of life's generativity from neoliberal productivity'. In turn, attempts to respond to smart objects tend to focus on either subverting smart objects in order to reclaim the spaces where these objects are having an effect, such as social networks, the cloud, smartphones and apps, or working alongside such technologies to generate new forms of political becoming.

Object politics

Distinct from responses to smart objects that are focused on the political compass of up and down, left and right, truth and power, collective and individual, the concept of phase reworks the distinction between up and down and left and right to focus on what Harman calls an object politics. This object politics draws upon, while also being distinct from an ANT approach to political issues. ANT attempts to dispute any politics that is based upon a fixed, *a priori* political position that is then applied to a particular issue. Rather, politics comes to be understood 'as successive moments in the trajectory of an issue' (Latour 2007: 812). In doing so, ANT focuses on the fragility of power. No one individual or technique is inherently powerful or 'has' power, because this power is only ever the effect of a network that needs to be continually remade and worked upon in order to exist. In Harman's (2014: 21) words, for Latour, 'the nature of the political always remains somewhat unknown'. As such, there is no inherent oppressor or oppressed group because power is potentially reversible at any moment. For instance, an actor in a network might stop working or create a new relation, which causes the network that produces the power, and thus the effect of this power, to break down. For Latour (2004b), politics is not about creating a revolution to tear down pre-existing structures. Instead it is about creating stronger or better networks that can be used to bring to light particular political problems as matters of concern 'whose import … will no longer be to debunk but to protect and to care' through the construction of further networks (Latour 2004b: 232). This means focusing on the process of politics, rather than any kind of outcome. As Harman (2014: 15) puts it, while political 'amateurs talk ends … professionals talk means'. For instance, rather than attempting to stop climate change on a global level through encouraging the uptake of renewable energy, an ANT perspective could be interested in identifying the networks that enable fossil fuels to appear as the default form of energy. Only once it was understood how fossil fuels appear as the default option could viable alternatives be suggested.

Although appealing, a network approach to politics is also problematic. As Harman argues, in order to maintain the idea that power is the effect of a network and thus reversible, Latour's (2004a) earlier work has to accept that every single entity within a network could be at least potentially political. Harman (2014: 56) terms this an 'ontologisation of the political'. An ontological politics leads to a flat, highly relational account of political agency where any entity is political because it might contribute to a network that enables one actor to control or shape others. While this seems enticing, the result is a position whereby the object itself only 'exists … [through] … other actors that register their presence' (Harman 2014: 58). In other words, Latour's account of political agency becomes something of a hall of mirrors. Objects are considered highly important nodes in helping to construct political power, but are also reduced to the way their presence is registered by other objects. Put in another way, power is considered to be the effects of various networked objects, but the power of these objects only has an effect when they shape or affect some other object in the network. The implication of this position is to ultimately downplay the specific capacities of particular objects, because objects themselves become spectral, being both the ultimate source of power and also powerless until they are enrolled into a network in a way that produces some capacity to shape or affect other bodies or objects.

In Latour's later work, in particular *An Inquiry into Modes of Existence* (2013), he attempts to develop an alternative to a flat ontological politics. Rather than all objects being potentially political in the same way, Latour argues that 'different types of reality require different standards of truth' (Harman 2014: 91). Latour (2013) sets out fifteen different 'ontological templates' that he uses to produce a pluralist account of how particular truths or statements become known through the creation of specific institutions, such as science, law and religion. Latour terms these templates 'modes'. Modes are 'types of continuity specific to each instance' that are based upon a set of particular values or veridications, or procedures through which the distinction between true and false is expressed (Latour 2013: 41–2). These veridications help generate prepositions. Here preposition literally means 'to mark a position taking that comes before a proposition is stated, determining how the proposition is to be grasped and thus constituting its interpretive key' (Latour 2013: 57). Prepositions are important because they 'play a decisive role in the understanding of what is to follow, by offering the type of relation needed to grasp the experience of the world in question' (Latour 2013: 57). In other words, prepositions structure how a particular issue or entity is experienced. For instance, a climate change denier would have a very different procedure to differentiate between what is true and false compared to a climate scientist. For the denier, climate science is not a valid procedure for measuring differences in climate temperature, whereas for the climate scientist temperature measurement is the very basis of any of their claims. These different procedures in turn might shape how each actor understands climate change as a political issue and how they would respond to it. For the denier, climate science could be considered a kind of conspiracy designed to hold back the growth of the economy, whereas for the climate scientist, climate science proves the urgency of radically cutting CO_2 emissions.

Through an account of modes, Latour recognizes the problems with purely relational theories of networks and politics. In particular Latour recognizes that studying networks as a series of relations soon becomes unsurprising. Every object or process of study shows the same thing; that the network is 'composed in a heterogeneous fashion of unexpected elements revealed by the investigation' (Latour 2013: 36). From this relational perspective, boundaries between distinct domains cannot be drawn and only exist as kinds of illusion for the actors who believe their domain to be distinct from others. However, Latour argues that such boundaries do exist. As networks are traced we immediately know that an issue is political, religious, a matter of law and the like. As such, 'there is thus a definition of boundary that does not depend on the notions of domain or network' (Latour 2013: 38). The truth and falseness of each mode are therefore contained within themselves, but also relational. For example, you can be upset about a court ruling from a religious perspective, while being satisfied that the legal procedure of the case has been fairly followed.

For Latour then, there is no general amorphous public that responds to political issues. Rather publics are constructed around issues that emerge from distinct domains that are mediated or disclosed by non-human things (also see Marres 2007). From this perspective, as Harman (2014: 162) suggests,

> The key move is to make all definitions of politics turn around the issues instead of having the issues enter into a ready-made political sphere to be dealt with. First define how things turn the public into a problem, and only then try to render more precise what is political, which procedures should be put into place, how the various assemblies reach closure, and so on.

In a similar manner to Latour and Harman, we could state that the phases of smart objects can produce publics by the way they disclose the qualities of objects through which issues become intelligible to people. In other words, a phase politics asks how particular smart objects disclose spaces and times that modulate the appearance of objects or actors as potentially political. As such, the politics of smart objects are not simply present in a situation, but emerge through particular issues that are disclosed by the relations between here and there, then and now that the perturbations of smart objects make intelligible. Formally defined, phase politics is a matter of interrogating how the phases of smart objects generate relations between near and far, here and there, then and now, which organizes the production of a public around an issue that is disclosed by a phase. From this perspective, politics is not about the political compass of up and down or left and right and does not involve arguing in terms of collective or individual good, or whether political decisions should be based on notions of truth or power. Instead, we could suggest that phases sensitize us to distinctions between what could be termed an endo and exo politics of smart objects.

Endo and exo politics

In a basic sense endo and exo mean either inside or outside, respectively. Biologists, for example, discuss different types of living beings as having either endo or exo skeletons. Most mammals, including humans have endoskeletons internal to the body, which work to support the muscles and internal organs. Crustaceans and arachnids, such as lobsters and spiders, have exoskeletons composed of chitin, with internal musculature attached to the exoskeleton by ingrowths called apodemes. Applying this notion of endo and exo to smart objects, by endo politics I mean politics that refer to how objects perturb one another to shape the production of a phase. By exo politics I mean how a phase modulates spatio-temporal intelligibility and shapes the references that humans use to make sense of their environment. A distinction between endo and exo modes of politics is important in order to analyse exactly how particular smart objects contribute to particular phases and in turn how these phases guide or shape the appearance of a political issue for a public. To make sense of the endo and exo politics of smart objects, let's turn to a specific example of cell phone towers in the village of Wishaw in the UK. Through an examination of these cell towers we can understand how smart objects, and the systems that enable these objects to generate phases, become political.

a) Endo politics

Think of the fears and concerns that have arisen around the use of smart objects such as smartphones in relation to human health. There are many recorded cases of people believing that they are negatively affected by radio waves and signals transmitted by smartphones and wireless devices, such as tablets and laptops (Wood 2006). These negative affects include conditions such as fatigue, depression and electromagnetic hypersensitivity. Electromagnetic hypersensitivity is a term used to describe a range of symptoms such as headaches and tiredness that people who are exposed to electromagnetic radiation are said to experience (Rubin et al. 2005). Electromagnetic hypersensitivity is a contentious condition because the majority of medical research on the topic argues that the radio waves generated by smartphones and the cell towers and Wi-Fi routers that smartphones use to connect to data services are not powerful enough to cause harm to humans (Drake 2006, 2010). As such, many medical professionals believe electromagnetic hypersensitivity to be a psychological ailment, rather than a physiological response caused by exposure to electromagnetic radiation (Rubin et al. 2008; Johansson et al. 2010).

Within the political compass of left and right and up and down, we might argue that the political issue of electromagnetic radiation in relation to human health is a matter of figuring out whether such hypersensitivities are real or not (truth politics), or in the case of cell phone towers specifically, a matter of fighting the corporations who place these towers in the environment (power politics). In turn this kind of politics could also be inflected with left and right tendencies. On the one hand, electromagnetic radiation could be understood as a public health concern (left), and on the other it could be positioned as a necessity for the healthy growth of a digitally mediated economy (right). Understood in endo and exo terms, such politics are quite different.

An endo politics focuses on how objects such as masts are arrayed or organized to generate electromagnetic radiation and how this radiation perturbs and discloses qualities of other objects, which creates the appearance of spatio-temporal relations within which electromagnetic radiation becomes a political issue. In relation to cell towers, endo politics involves the particular procedures, techniques and components used in the design of cell towers and the planning of rules and regulations that dictate how towers are erected. For instance, in the UK there are approximately 52,500 cell towers, technically called base stations, which are used to transmit mobile phone signals (Mobile Mast 2013). These base stations are supplemented by various 'cells'; smaller antennas that provide infill coverage necessary for signals to travel. Macrocells provide the main coverage for mobile networks and are positioned at a height that exceeds surrounding structures and topography. Microcells are smaller antenna mounted lower to the ground and provide additional capacity to the network where there are many mobile users. Finally, picocells are used to boost signal strength inside buildings and other areas, where the signal strength from the macro and micro cells are weak.

For telecommunication companies and the UK government, the politics of phone masts are orientated around the ability of the masts to perturb and be perturbed by radio waves and enable communication to take place. In order for mobile networks to operate continuously, the signals from various stations and cells need to overlap, which in turn partly dictates the decision-making processes regarding the location of base stations. Of course, telecommunication companies cannot simply place a cell or base station wherever they like. In the UK, planning decisions around mobile phone masts are based upon the National Planning Policy Framework (NPPF). When providing permission for the erection of a mast, the NPPF (2012: 11) document states that local councils must balance the need for an 'expansion of electronic communications networks' while aiming 'to keep the numbers of radio and telecommunications masts and the sites for such installations to a minimum consistent with the efficient operation of the network'.

The spatial politics of these masts are understood and communicated through quantitative spatial measures, such as how many metres masts have to be located from the ground, or the distance in metres between masts. For instance, the UK Planning Policy Guidance on Telecommunications

(PPGT) states that approval is required for the construction of the following types of mast:

> (i) a ground based mast of up to and including 15 metres in height; (ii) a mast of up to and including 15m in height installed on a building or structure; (iii) an antenna (including any supporting structure) which exceeds the height of the building or structure (other than a mast) by 4 metres or more at the point where it is installed or to be installed. (DCLC, 2001: n.p.)

From an endo political perspective, these documents are not important because they dictate the metrical size or height of these objects or where they can be placed 'in' space or time. Nor can these documents be reduced to a left- and down-focused analysis that would understand them as techniques that enable space to be partitioned and divided according to the commercial logics of private telecommunication companies, who erect the masts in the pursuit of profit. Rather, these documents matter because they dictate the conditions of possibility for generating the appearance of spatio-temporal relations where masts appear as distinct objects and so form part of a particular diffuse or partitioned zone, within which the effects of these masts are presumed to exist. In other words, it is not the height or size of the masts per se that matter, but how they modulate the appearance of near and far and then and now because it is these appearances that shape whether people consider themselves to be perturbed by these masts.

b) Exo Politics

The endo politics of documents like the NPPF generates a set of exo political effects in the sense that they disclose entities as spatio-temporal issues, which in turn shape the formation of a political public around the issues these entities raise. We can theorize political publics as formed by a set of shared practical, affective and emotional references. In relation to the political publics formed around phone masts, take for example the small village of Wishaw in Warwickshire in the UK. Between 2001 and 2003 Wishaw was at the centre of fears around the health dangers associated with mobile phone masts. In 1994, a 22 metre mast had been installed in the village and in the intervening period residents complained that it had caused a marked rise in ill health (Foggo 2003). People living in vicinity of the mast pointed to a whole host of symptoms that had supposedly begun since its erection, including headaches, skin rashes and a rise in cancer. In one resident's words, the simple presence of the tower and the microwave radiation it transmitted and received felt 'like a pneumatic drill going outside your house' (Out 2004: n.p.). The resident directly linked the mast to their illness, stating 'you can't hear it but your body cells are being impacted by this pulsing microwave radiation' (Out 2004: n.p.). Public outcry and protest by the residents of Wishaw resulted in the mast's destruction by vandals in 2003 and a promise by T-Mobile (the mobile telecommunications company that owned the mast), not to replace it.

We could argue that the original T-Mobile mast installed in 1994 was the key to generating phase spaces and times by disclosing relations between here and there and then and now as it perturbed and was perturbed by the residents who lived in the village. Primarily, relations of here and there, then and now emerged through the physical visibility and presence of the object of the mast. At 22 metres tall and located in the middle of flat farm land surrounded by roads and houses, the mast was clearly visible throughout the village and served as a prominent landmark. In turn, this visibility worked to denote the presence of electromagnetic radiation in the area by creating relations of distance and proximity. The very sight of the mast created a sense of proximity to the radiation itself. Visibility of the mast became a reference to the presence of electromagnetic radiation that was not visible to the eye, which in turn contributed to the intelligibility of the phase, within which the effects of the radiation were assumed to exist.

In Chapter 3, we discussed the concept of reference in relation to Heidegger as a kind of human practical usability. We developed the concept of reference to distinguish between the way humans assign references to objects as they use them, with the concept of phase, where smart objects perturb one another and generate references outside of a relation with human beings. In the context of exo politics, we can consider how phases generate references between objects outside of human intelligibility, but can also influence the kinds of references humans attach to these objects. In the same way, we can also investigate how humans attach references to smart objects in ways that indirectly overlap with how smart objects perturb and are perturbed by other objects.

Expanding Heidegger's account of human references, we can state that references are not just forms of practical usability, but can also be affective and emotional as well. Understanding references as affective and emotional is the key to analysing the exo politics of phases. While there are many different theories of affect, for our purposes affect can be defined as a form of pre-personal bodily intensity (Massumi 2002; Anderson 2006; Pedwell 2014). Distinct from affects, emotions are how humans make sense of affects in relation to their particular embodied, historical situation and biography (Gregg and Seigworth 2010; Wetherell 2012; Coleman 2015). For example, in relation to media, an affect might be the rising of hair on the back of the neck when watching a horror film or viewing a GIF animation (Ash 2015b), which will be interpreted differently depending on the viewer's previous experience. Some viewers will experience the affect of the film or GIF as an emotion of fear, whereas others will make sense of a similar affect as an emotion of excitement or pleasure. The spaces that phases make intelligible can be understood as affective in the sense that they can create particular bodily responses that are experienced through emotional registers that are shared and communicated between groups of people. Affective and emotional references are therefore the exo references between the residents' bodies and the smart objects and other technical objects that made up the phase, which are shaped by the endo perturbations between the mast and radio waves the mast utilizes to perturb other objects, such as mobile phones.

The idea that references are affective and emotional as well as practical is useful in order to show how the spatio-temporal intelligibility generated by the phase of the Wishaw mast altered how the mast appeared as a distinct political issue. The affective and emotional nature of references directly fed into the kinds of region that appeared for the residents. For example, as local resistance to the mast gathered strength, residents formed a local action group: Seriously Concerned Residents Against Masts (SCRAM), which held public meetings and connected with other national and international organizations that were lobbying against microwave radiation (O'Connor 2007). We could state that this resistance served to reinforce the durability of the phase of the mast by creating ever more complex and far reaching affective and emotional references between the residents of the village and the mast itself.

The spaces and times disclosed by the phase of the mast were further reinforced by residents who recorded cases of cancer and their proximity to the mast through adding incidents to an Ordinance Survey map of the village. This map was then used as evidence to reinforce the residents' claims in public consultation meetings regarding the mast (O'Connor 2007). The Ordinance Survey map created and reinforced the existing references between the mast and villagers by creating a durable object that visualized the 'hidden effects' of the mast and displayed the spatial proximity of these effects to the mast. But even with the use of maps, the phase the mast produced was not necessarily experienced metrically. Rather, distance became intelligible through a series of affective and emotional references. In this sense, the presumed ability of the mast to perturb the bodies of the residents shaped an experience or understanding of distance. The further the residents were from the mast, the smaller it appeared, which reduced the potency of the references that the phase of the mast disclosed to the residents.

Other objects the residents used to attempt to prove the problematic nature of the phase only served to intensify the references they constructed in relation to the mast. For instance, alongside travelling to Westminster to speak to parliament, some residents purchased MW1 microwave electro-smog detectors (O'Connor 2007). These detectors are perturbed by microwave radiation and disclose qualities of this radiation as sound. They are specifically marketed at individuals concerned with the potential health effects of exposure to electromagnetic radiation. Sensory Perspective (2012: n.p.), the manufacturer of the MW1 describes their product in the following way:

> This new 'broadband' detector exposes and converts the electromagnetic impulses it encounters into a collection of sounds (buzzing, screeching, pulsing) that is far more varied and illuminating than the 'click' of a geiger counter. You can hear the amount, type, amplitude, and quality of the pulsed electronic pollution created by local transmitters/emitters operating between 50MHz and 3000 MHz, and enjoy the silence when the environment is clear. Hearing is believing.

The phrase 'hearing is believing' is the key here. The MW1 provides no quantitative read out of the types or metrical intensity of microwave radiation present in an environment. The manufacturers also insist that the MW1 'is not an alarm, nor is it a warning device' (Sensory Perspective, 2012: n.p.). From an exo political perspective, the MW1 detector works to alter the intelligibility of the phase, modulating the appearance of the phase from a vague, diffuse region to one that envelops and surrounds the user. The MW1 achieves this through the way it is perturbed by the electromagnetic radiation and discloses qualities of the radiation as a sound wave emitted from its speaker. In doing so the invisible radiation becomes an audible sound that can be registered by human sensory perception.

As Simpson (2009, 2015), Kanngeiser (2012) and Thomson (2017) argue, hearing unlike other senses such as vision infiltrates the body and cannot be easily ignored. As a resonant object, sound has the potential to activate a very raw and immediate set of affects. In Gallagher's (2016: 44) words,

> What begins as a flow of raw vibration may produce sensations, emotions or moods, or push through into the realm of significance to be heard as anything from slight hints of something, evoked memories, associations or senses of space, through to more formal meanings and representations, as in spoken language. Sonic affects may accumulate layers of significance over time, through repetition and habit, by becoming attached to other affects.

Using sounds such as 'buzzing, screeching and pulsing' (Sensory Perspective 2012: n.p.) as a medium for the expression of the presence of microwave radiation encourages an immediate affective bodily response to the MW1 device. Already primed with negative references to the mast, the MW1 could encourage users to experience these affects in a negative way through emotions such as fear or anxiety, without recourse to any kind of analysis of context. Only when the MW1 is silent does the user know that microwaves are absent in the environment and they are 'safe'. Devices such as the MW1 generate further affective and emotional references by multiplying the mediums through which negative references are made. The screeching sounds from the MW1 create an oral unpleasantness that works to confirm and reinforce other unpleasant symptoms that some residents of Wishaw were experiencing such as headaches and skin rashes. Through this process, the MW1 modulates how space and time appear to the user. Rather than a diffuse zone where nearness and distance were experienced by the visual proximity to the mast, residents using the MW1 became more sensitized to the phase of the mast, as the MW1 produced an envelopment. The screeching of the MW1 intensified the phase generated by the mast, creating the impression that the perturbations of the electromagnetic radiation and thus the mast were close by, which perhaps only served to further intensify the residents' anxiety and worry about the effects of the mast.

What is important to note here is that phases can generate all manner of exo references that are real and have effects, even when these references are not equivalent to the actual endo perturbations between smart and technical objects and human bodies. From the perspective of phase politics, it doesn't matter if the electromagnetic radiation from the mast in Wishaw does affect human health. What does matter is that the residents consider the phase to have exo effects, which in turn shapes the kinds of references that are constructed between themselves and the mast. In this case, the endo effects of the phase of the mast are real and do affect human health, but not necessarily in the way participants who suffer from these health problems consider.

Phase politics thus emerge at the intersection of these different modes of endo and exo references. As the example of Wishaw suggests, phase politics is not just a matter of campaigning for or against a particular issue. But neither is politics simply about the production of a public that emerges around a particular issue as Latour and Harman suggest. Distinct from these accounts, phase politics points out that political issues around smart objects only appear to anyone or anything through the way phases shape the appearance of space and time and it is through this space and time that particular entities become intelligible and referenced within a human structure of affective and emotional concern.

In Wishaw the issue of the phone mast became a matter of concern because of the references disclosed by the phase of the mast and the electromagnetic radiation that perturbed other objects, which residents linked to their health. The references associated with the phase were aided by a range of technical objects, such as maps and the MW1 electro-smog detector. If the mast had not been so intelligible as an object (perhaps 5 metres from the ground rather than 22), would the residents have created the original reference between the mast and their health? In other words, the phases objects generate act as the conditions of possibility for shaping the references around which a political response is organized based on the compass of up and down, left and right that I introduced earlier in the chapter. It is only when the mast constructs a phase, through which references between objects can be made, that one can take a position of left or right, truth or power. SCRAM emerged in response to the phase of the mast as a left leaning political organization in the sense they opposed the imposition of the mast by a private corporation and the landowners who had agreed to host the mast on their land. At the same time SCRAM were concerned with truth politics in the sense that they were looking for recognition that electromagnetic radiation can cause serious health problems in order to validate their wish to have the mast removed from the area. To summarize, phase politics is not about judging who is right or wrong in a situation. Instead it is about understanding how certain issues around smart objects appear via a phase. The phases of smart objects can then shape the spatio-temporal intelligibility of the objects in that phase and in doing so create a public that is affectively and emotionally linked to those objects.

Phase politics

The approach to phase politics developed in this chapter is distinct from current new media theory that considers smart objects to be a logical extension of neoliberal forms of control. At the same time, phase politics are also distinct from Latour's network politics or his later work on the modes of existence. Rather than being defined purely through their relational connections with other things as part of a network, or as a specific mode of existence, smart objects have an autonomous reality whose perturbations generate phases and disclose spatio-temporal relations between objects. It is my contention that through phases and the spatio-temporal relations they disclose, particular publics form around a specific issue in relation to smart objects. From this perspective there is no public before or prior to the appearance of an issue that is produced through a phase. In Harman's (2014: 172) words, 'each issue/object generates a new public, instead of the same grey anonymous mass weighing in foolishly on every possible topic'.

Understanding politics in this manner guides us away from accounts that read smart objects as primarily expressing a logical continuation of a broader neoliberal project. While not denying that neoliberal processes or logics may influence the development and employment of smart objects, phase politics does not assume that the politics of smart objects is necessarily or inevitably linked to neoliberal practices. Instead, phase politics is far more concerned with the issues particular smart objects create, despite of neoliberalism. Of course, the case of the Wishaw phone mast is only one example of the politics of phases. The same concepts of endo and exo politics could be applied to analyse how the phases of smart objects contribute to the production of publics in a variety of settings. On a more grandiose scale, one could analyse the role that smart objects such as smartphones played in the Arab Spring in order to produce a public around the issue of democratic representation (Khondker 2011). On a smaller scale, a phase politics could be used to understand how smart objects contribute to the production of publics around energy usage, through the adoption of smart energy meters (Strengers 2013).

In this chapter I have outlined how phases organize a public around an issue. However, I have not suggested how publics could respond to the concerns that phases make intelligible. The next chapter turns to this issue in detail by developing the concept of involution.

6

Involution

In *The Limits of Critique*, Felski suggests that academic criticism has become an ineffectual mode of engagement. Too often critique becomes mechanical, a kind of well-intentioned platitude that 'responds more or less the same way to any and all new events' and 'treats … works as objects both possessing inherent value and requiring systematic suspicion' (Felski 2015: 3). Here critique 'prime[s] us to approach [objects] in a given state of mind, to adopt a certain attitude toward our object' (Felski 2015: 20). Unfortunately, the very act of treating an object with a suspicious attitude often mitigates a close analysis of the object itself. In turn, suspicion becomes both the expression of critique and the moment when academic and political intervention begins and ends. The form of critique Felski (2015: 52) describes is based upon two techniques that she terms 'digging down' and 'standing back'. In digging down critics act as kinds of inspectors or detectives, sifting through objects in an attempt to unveil the hidden processes that structure a situation and thus reveal the truth of social reality. A process of standing back often accompanies this digging down, where critics position themselves, either implicitly or explicitly, as separate or detached from the social reality they are critiquing. Through this manoeuvre, critics attempt to create the appearance of impartiality in order to strengthen the basis of their claims.

The main issue with this form of critique is that it can result in the practice of endless contextualization.

> In diagnosing the insufficiencies of a work … critique explains these insufficiencies by invoking some larger frame. It looks behind the text for some final explanation or cause: social, cultural, psychoanalytical, historical or linguistic. The text is derived, in a fundamental sense, from something else. Critique itself, however, remains the ultimate horizon – it is not an object to be contextualised but is itself the ultimate context. (Felski 2015: 189–90)

In doing so, critique becomes a kind of constant deferral. The particular object under analysis tends to be explained away through recourse to some general theory such as Marxism or Functionalism or other framing concept such as society or culture.

Opposed to the act of inspection, detection and contextualization, Felski proposes that social scientists compose. Felski (2015: 182) defines composition as 'creative remaking – that binds … [people] … in ongoing struggles, translations and negotiations'. Thinking compositionally is a matter of 'addition rather than subtraction, translation rather than separation, connection rather than isolation, composition rather than critique' (Felski 2015: 182). Felski works within the domain of literature and literary theory, where composition is a matter of generating new readings and understandings of texts. However, her notion of composition is useful for thinking about smart objects. In the previous chapter we discussed what I termed phase politics and how the spatio-temporal relations phases disclose shaped peoples' understanding and references to things. Following Felski, if we want to alter the kinds of phases that emerge and the references and regions phases attempt to generate, then a simple critique of phases as a force of venture capitalism or neoliberalism is not enough. Instead, we can develop new modes of composition that are sensitive to smart objects and phases. If phases are constructed through the qualities disclosed through the perturbations between smart objects, then we can alter phases by changing the way smart objects perturb one another, which shapes the qualities and thus spatio-temporal relations they disclose. We can term this process of composition involution.

To define and make sense of the concept of involution, the rest of the chapter forms four sections. The section 'Involution' turns to the work of Catherine Malabou and Gilles Deleuze to define involution. The section 'Struction and Dis-struction' uses the work of Nancy and Barrau to argue that there are two main forms of involution: structive and dis-structive involution and works to unpack this distinction. The section 'Structive Involution' focuses on structive involution and uses the example of the ongoing negotiations between Uber and Transport for London as a competition over the construction and involution of phases. The section 'Dis-structive Involution' turns to the example of the Hong Kong Umbrella protests in 2014 to understand dis-structive involution and specifically how phases can be structed to dismantle and reorganize the spaces and times smart objects make intelligible. The final section 'Phase Activism' turns to the notion of phase activism and demonstrates how the different modes of involution discussed in the chapter can be used to actively question, unpack and rework phases that might be detrimental to the appearance of issues as potentially political. Understanding how, where and why involutions occur is a case of identifying various mechanisms that shape the collapse and differentiation of phases into new forms. By making such involutions explicit it is hoped that the chapter will provide tools for others to compose involutions in relation to issues that they recognize as political.

Involution

Involution is a complex term with a variety of meanings. From a mathematical perspective the term refers to a construction in the differential geometry of curves (Lachlan 1893). More simply put, involution is a process whereby something turns in on itself, such as folding a piece of paper in half, causing the separate edges of the surface to come into contact with one another. The term involution has also been developed as a critical concept. For Malabou (2002, 2004), involution is related to her theory of plasticity and in particular a notion of time as plastic. As Crockett (2012: xxvii) suggests in relation to Malabou's work, 'Time is plasticity itself, absolute plasticity . . . . The form of plastic time is bifurcation, which leads to a fractalizing of temporalization, an unfathomable involution.' In Malabou's reading, involution is a process of bifurcation, where an object changes through internal processes of differentiation.

Distinct from Malabou, Deleuze and Guattari discuss the term in relation to notions of biological evolution and change. For Deleuze and Guattari (1987), the concept of involution is an attempt to think the emergence of difference as shaped by molecular forces in the environment, rather than molar forces of genetics that are biologically inherited through reproduction and natural selection. Focusing on molecular forces and affects, Deleuze and Guattari question a model of life as predicated on a bounded organism (Hansen 2000a). According to Gontarski (2012: 238), for Deleuze and Guattari, involution 'is in no way confused with regression'. More specifically, involution 'does not go from something less differentiated to something more differentiated' (Deleuze and Guattari 1987: 238). Rather it forms a block 'that runs its own line "between" the terms in play and beneath assignable relations' (Deleuze and Guattari 1987: 239). Here a biological organism is not defined 'by its characteristics (specific, generic, etc.) but by populations that vary from milieu to milieu, or within the same milieu; movement occurs not only, or not primarily, by filiative productions but also by transversal communications by heterogeneous populations' (Deleuze and Guattari 1987: 239). For instance, for Deleuze and Guattari a pig is less of a biological organism from the genus 'Sus' and more of a composition of forces that emerge from an intersection of other things such as truffles, trees, mud, bacteria and so on.

While very different from one another, what is interesting about Malabou and Deleuze and Guattari's use of the term involution is that both suggest that involution is not a matter of explosion or implosion. Involution is instead more of an active process of becoming through which capacities and differences are generated and propagated. Developing this point, in relation to phases, involution can be defined as the active modification of smart objects that make up a phase in order to shift or change the intelligibility of the spaces and times that the phase discloses. In other words, involution is a process through which spatial and temporal intelligibility is actively recomposed. Involutions are compositional in the sense that they attempt to generate new possibilities from the very objects that make up the phase. Rather than a moment in which smart objects are reduced to 'some final explanation or caus[al] context' (Felski 2015: 189), involution is a matter of disrupting and modifying phases through a series of active processes of what I term struction and dis-struction.

Struction and dis-struction

The term struction emerges from Nancy's work with the physicist Barrau (see Nancy and Barrau 2015). Distinct from the idea of composition, which Felski develops to speak to the academic discipline of literary criticism, Nancy and Barrau's language of struction provides the basis for a more focused vocabulary that can be developed to discuss smart objects and phases specifically. Nancy and Barrau advance the concept of struction to provide an image of making that is distinct from logics of construction that have come to dominate the modern world. They position struction as occupying a conceptual middle road between the idea of construction on the one hand and deconstruction on the other. In Nancy and Barrau's (2015: 2) words, 'Construction and deconstruction are closely interconnected with one another. What is constructed according to a logic of ends and means is deconstructed when it comes into contact with the outermost edge where ends reveal themselves to be endless and where means, for their part, reveal themselves to be temporary ends that generate new possibilities for construction.' Construction therefore refers to literal processes of making, where bridges are engineered and built, railway tracks are laid and so on. Deconstruction refers to techniques of thought for questioning and unpacking the internal and relational qualities of systems. In the work of Derrida (1998) for example, deconstruction is a matter of analysing texts in order to expose the inherent instability of meaning that enables that text to make sense (Howells 2013).

Distinct from either construction or deconstruction, struction 'signifies "to amass," "to heap." It is truly not a question of order or organization that is implied by con- and in-struction. It is the heap, the non-assembled ensemble. Surely it is contiguity and co-presence, but without a principle of coordination' (Nancy and Barrau 2015: 4). Nancy and Barrau consider struction as both the organizing principle of technologically mediated life and a habit of thinking that provide opportunities to create new worlds. For our purposes struction can be defined as a form of making that does not attempt to create a total or complete product or come to a distinct resolution. Following this definition, we can state that two types of involution are possible: structive involution and dis-structive involution.

Structive involution recognizes that phases are kinds of non-assembled ensembles in the sense that phases are shaped by all manner of objects that exist alongside smart objects, which both perturb and fail to perturb entities that make up the phase. Formally defined, structive involution refers to how individuals can alter or modify objects in general and smart objects in particular to change the intelligibility of space and time that appear through these objects. Structive involution is therefore not a simple case of construction. Whereas construction is based upon the notion of a plan or project that the process of construction seeks to realize, struction is a matter of placing objects alongside one another in ways that disclose qualities that alter the appearance of the spaces and times of the phase, but whose effects can never be known or anticipated fully in advance. As we will explore in the following two sections, different forms of struction can involve modifying and dismantling objects on a number of levels. Structions might involve simply using pre-existing functions of smart objects in a new or original way, modifying smart objects to enable new functions, or altering the legislation around the ways that smart objects can be used or employed.

Dis-structive involution refers to the way that smart objects can be used to dismantle the space-times and objects that phases makes intelligible. Like the difference between struction and construction, it is important to note the distinction between what I am calling dis-struction and the everyday term destruction. In common use, destruction refers to the process of causing damage to an object to the point where it cannot fulfil its primary function or be repaired. When destroying, you knock over, pull down, tear up, raze, level or wreck. Here objects are altered to the extent that they are no longer recognizable or no longer fulfil their primary function. The term dis-struction, is much closer to a notion of dismantling and re-mantling. Dismantling refers to a deliberate, often more careful process of taking something apart without destroying its key components. When destroying a building, one might use a wrecking ball to smash it to the ground. But, when dismantling a building one might take it apart brick by brick in order to salvage and reuse the materials in a different project or to simply heap them together. Therefore, dis-struction is a matter of undoing rather than simply destroying or wrecking, as in processes of destruction. To make sense of these two somewhat abstract terms of struction and dis-struction, let's turn to examples of structive and dis-structive involution in action: the Uber smartphone app in London and the Firechat smartphone app in Hong Kong.

Structive involution

Uber is a 'ride-sharing' or 'ride hailing' company that is organized around an app that is available for iOS and Android devices. Activating the app, users can request a ride without calling a centralized taxi company, and payment is automatically taken from a credit or debit card kept on file within the app. In Uber's (2016) words, 'Uber is the smartest way to get around. One tap and a car comes directly to you. Your driver knows exactly where to go. And payment is completely cashless.' To enable this service, the Uber app mediates both ends of the transaction. When users activate the app they are given an estimate of when their ride will arrive. When they request a ride, nearby Uber drivers are alerted on their app and can accept the ride. Once they have picked up a passenger, the Uber app then provides the driver with step-by-step directions to the passenger's destination and upon completion the passenger's card is automatically debited and sent to the driver's bank account, minus Uber's fee.

Uber emphasizes speed and simplicity as the key to its competitive advantage over conventional taxis. Drivers can work as much or as little as they like and Uber claims passengers save money compared to using a regular taxi. Uber's business model is based around the eradication of what is known in the taxi industry as search costs. Search costs are the time and money it takes for drivers to locate a fare.

> Rather than calling a dispatcher and waiting, or standing on the street, users can hail a car from indoors and watch its progress toward their location. Drivers also cannot poach one another's pre-committed fares. This is a real boon for consumers who don't like long waits or uncertainty – which is to say everyone. Uber can also advise drivers on when to enter and exit the market – for example, by encouraging part-time drivers to work a few hours on weekend nights. (Rogers 2015: 88)

Alongside its convenience, Uber's success is partly based on how its app constructs a phase and the way this phase modulates the spatio-temporal intelligibility of Uber's users and drivers. The Uber app modulates the spatio-temporal intelligibility of users in a number of ways. Upon opening the app a black pin denotes the user's location on a map with a black notice above stating, 'SET PICKUP LOCATION'. On the left side of this notice is a circle with a dot that moves clockwise around its edge. Within the circle is a time such as '5 MIN'. This denotes the estimated time it would take for a car to arrive if you booked the ride. Alongside this estimation a number of Uber car icons move around the streets in the vicinity of the user's location. However, these icons do not exactly replicate where Uber vehicles really are. In Hwang and Elish's (2015: n.p.) words:

> The presence of those virtual cars on the passenger's screen does not necessarily reflect an accurate number of drivers who are physically present or their precise locations. Instead, these phantom cars are part of a 'visual effect' that Uber uses to emphasize the proximity of drivers to passengers. Not surprisingly, the visual effect shows cars nearby, even when they might not actually exist. Demand, in this case, sees a simulated picture of supply. Whether you are a driver or a rider, the algorithm operating behind the curtain at Uber shows a through-the-looking-glass version of supply and demand.

Together these visual and textual icons work to create a sense of spatial and temporal immediacy – an appearance that the app is providing real-time spatial updates of the availability and location of its drivers, even when it is not. Utilizing these simple techniques, the Uber app modulates the spatio-temporal intelligibility of users around a logic of envelopment. The effect of this envelopment is the creation of an impression for Uber users that Uber cars are nearer rather than further away and available closer to now rather than later. In turn, the Uber phase also creates the appearance that there are a large number of available cars waiting to be hired. It would appear that Uber utilizes this technique because the more proximate the spatial and temporal relations of Uber cars appear to be, the more likely it is for the user to request a ride. In this case, modulation is designed to stimulate demand through the disclosure of relations of near and far and then and now between the Uber passenger and the Uber vehicles that appear via the app.

Generating a phase organized around the logic of envelopment is not only used to try and encourage ride requests. Drivers are also subject to similar techniques that attempt to encourage them to move to particular locations to look for fares. Uber achieves this through the use of surge zones that it displays on drivers' smartphones (these zones are not visible on passengers' apps). The surge feature of Uber refers to an increased price that passengers have to pay during 'busy periods'. Surge pricing can increase a fare anywhere between 1.2 and 4.5 times the cost of the base rate fee. As Hwang and Elish (2015: n.p.) again point out,

> Drivers are shown a map of 'surge zones', which ostensibly reflect the demand for rides in different parts of the city at a given time. While this is how the company frames it, it actually isn't the case in practice. According to its patents, Uber generates surge based on the projected demand of riders at some point in the future. When it works, this system produces low latency – a rider requesting a car can get one quickly. When it doesn't, drivers can spend precious time and gas in a neighborhood with no or slow demand. The suppliers get to see only what a system expects the state of the market to be, and not the market itself.

The surge zone map is organized around a logic of partition. The map splits the city or area the driver is working in into different coloured regions. 'Red means surge is in effect, orange means demand is building, yellow indicates demand is there' (Rosenblat and Stark 2015: 7). These zones can be quite effective in shaping driver intelligibility and in turn behaviour. As Rosenblat and Stark

(2015: 7) suggest, 'Some drivers refer to surge as a "herding tool" that ushers them into specific geo-fences.' The surge pricing algorithm thus works to generate a phase that modulates the spatial and temporal intelligibility of the driver. By anticipating demand, the app generates spatial and temporal relations of proximity and distance that the driver associates with a greater chance of getting a fare. Depending on their current location they might travel to where surge is already in effect or wait in an orange zone, on the assumption that the surge pricing will soon activate, giving them the first chance of receiving surge fares in that zone. The surge pricing also modulates the spatio-temporal intelligibility on the passenger side of the transaction around a spatial logic of diffusion. Upon opening the app, the passenger does not know that surge pricing is in effect until they input their destination and call the ride. The app then presents the user with the surge pricing charge and asks them if they are willing to accept this charge. While the space disclosed by the phase is clearly defined for drivers (albeit disingenuous), the space and time disclosed for passengers is deliberately vague and open. With no way of knowing the spatial or temporal 'edge' of the surge region, Uber passengers are encouraged to accept the higher fare because they have no way of navigating to a non-surge area.

Drawing upon these two examples, it is clear that the Uber app is designed to produce a phase that modulates the spatio-temporal intelligibility of both Uber customers and drivers. These phases and their associated intelligibilities are a central part of Uber's business practice to both stimulate consumer demand and attempt to encourage its drivers' to drive when and where Uber want them to. In light of these practices, Uber has been heavily criticized. Beyond the way data are presented to drivers and passengers, Rogers (2015) identifies six other areas where Uber's policies are considered to be problematic.

> First … [is the claim that Uber] … is unfairly competing with taxi drivers by entering their market without following regulations or fare schedules;

> second, that it aspires to become a monopoly; third, that its cars or drivers are unsafe or underinsured; fourth, that it may invade customers' privacy; fifth, that it enables discrimination by drivers and passengers; and sixth, that it is undermining working standards for taxi drivers and compensating its own drivers poorly. (Rogers 2015: 91)

With these criticisms and issues in mind, governments and officials in a number of countries have attempted to regulate how Uber operates. Some of these regulations have involved outright bans on the service, such as in India and Thailand (Khosla 2015). In Germany, regulators have demanded that drivers must be licensed before they can drive for Uber, resulting in the withdrawal of its cheaper Uberpop service (Che 2015). As part of this regulatory response, the governing body for public transport in London in the UK has recently engaged in consultations around the way Uber can operate in the city. Through a range of research with public and professional bodies, Transport for London has produced a consultation document that provides recommendations about the ways Uber should be allowed to work in London. Rather than recommending an outright ban, Transport for London have instead asked for some more minor changes that are linked to how the app produces a phase. For example, the Transport for London (2016: 10) consultation document states:

> Operators must not show vehicles being available for immediate hire, either visibly or virtually via an app. A number of consultees have suggested that, whether through an app or through physical street ranking, some operators are creating the impression of vehicles being available for immediate hire. This is increasing the risk of unauthorised vehicle/driver 'touting' and other illegal cab activity. Operators with a physical base for their drivers (e.g. a local minicab office) could still have vehicles at the premises awaiting a booked journey.

By forcing Uber to alter the way it presents cars in the app, Transport for London is attempting to modify the objects that make up the phase in order to alter how space and time appears for Uber users. Rather than a space and time organized around a logic of envelopment and gradation, where Uber cars appear to be nearby at all times, removing the car icons from the app creates a sense of spatial and temporal distance and imply that Uber vehicles are not close to the user. Transport for London justifies this change because it considers the car icons to encourage 'the risk of unauthorised vehicle/driver "touting" and other illegal cab activity'. However, it does not explain why it considers this to be the case. As any Uber user knows, once you have booked an Uber, the car's make, colour and registration is visible in the app, which actively discourages the chance of using an illegal or unregulated vehicle. Instead, it appears that Transport for London is actually advocating a change in how the app works in order to initiate an involution of the phase. This involution involves altering the objects that appear within the app and thus modulating the spatio-temporal intelligibility of users in ways that do not benefit Uber's commercial strategy (that seeks to maximize the number of passengers that use Uber). In other words by altering what appears in the app, the Uber phase becomes structively re-organzed around a spatial logic of diffusion rather than envelopment and a temporal logic of dilation rather than gradation. Of course, Transport for London would not use such language, but it is pushing for a change in how the app internally differentiates between objects, which in turn alters how space and time appear for the user of the app. In the same way that Uber simulates car icons in the app to encourage the user to book an Uber, it would appear that Transport for London wants to remove these icons in order to discourage the user from booking an Uber and use a licensed black cab instead.

Just as Uber constructs a phase for both the Uber driver and passenger, Transport for London is also attempting to involute the phase on the Uber driver's side as well. This is evidenced by an increase in the minimum topographical skills that London Uber drivers will need to demonstrate to be able to carry fares in the city. As the consultancy document states:

> We shall introduce a significant enhancement in the content, management and delivery of the PHV Topographical Skills Assessment (TSA or 'the test') and an applicant's general understanding of the rules that govern private hire licensing. We will invigilate the TSA going forward, utilising a core of the existing test centres. The new test would be delivered in three parts: A computer package testing map reading skills; The ability to navigate to/from key points in London (e.g. mainline stations); Clear understanding of private hire licensing regulations. We will not provide training for the enhanced TSA and instead would encourage candidates to attend accredited operator training centres to adequately prepare for the test. (TFL 2016: 22)

This proposed change in regulation is clearly designed to decrease the ease with which people can become Uber drivers. Part of Uber's appeal to drivers is that the app takes care of all parts of the transaction with the customer. The app has turn-by-turn GPS navigation built into its core functionality, meaning that the app leads the driver to where the passenger is and then leads the driver to the passenger's destination. Point-to-point navigation means that Uber drivers can successfully work with no navigational knowledge of the city or area they are driving in. By increasing the difficulty of the TSA test, Transport for London is attempting to create an obstacle that stops drivers from accessing the Uber app in the first place. When using turn-by-turn navigation in the Uber app, the driver's spatio-temporal intelligibility is organized by a logic of envelopment and dilation. This envelopment and dilation is evidenced by the audio and visual references that tell the driver when and where to turn. In doing so, there is no need to relate their current location to particular landmarks, or to road names. Here the spatial and temporal relations between near and far, then and now emerge through a relationship with the stationary smartphone screen located in the driver's car, rather than in relation to objects that the driver moves past. In turn, when experienced through the Uber app, the city of London perhaps appears as a set of continuously near objects to the driver (the green arrows on screen that tell the driver when and where to turn), creating a spatio-temporal intelligibility based around continuing spatial and temporal proximity between the driver and their vehicle.

Transport for London's intervention is then a pre-emptive kind of structive involution in the sense that it attempts to stop Uber from being able to construct a phase between its app and the driver. If the driver is not willing to spend the time and energy to internalize information about London then they cannot be licensed and therefore do not have the opportunity to access the enveloped and dilated spatio-temporal intelligibility the phase of the app discloses to the driver, which they rely upon for way finding. In other words, Transport for London is attempting to create an involution by altering how spaces and times appear in the app by changing the rules that govern whether one can access the phase of the app and its associated intelligibilities. However, not all involutions attempt to structively reorganize the logic of how space and time appears. While structive involution works with phases to alter the intelligibility of space and time, dis-structive involution attempts to dismantle the objects, spaces and times that are disclosed via phases. To explain dis-structive involution we can turn to the example of the Umbrella movement in Hong Kong.

Dis-structive involution

The Umbrella revolution, or Umbrella movement, is the name given to a series of sit-in street protests of public spaces in Hong Kong. These protests began on 26th September 2014 and lasted for 80 days. The protests were a response to China's desire to pre-screen candidates who were able to stand in Hong Kong's elections under the one-country, two-party system that was agreed when Hong

Kong was handed back to Chinese control by the British in 1997. According to Lee et al. (2016: 356), this event was called 'the Umbrella Movement by ... [non-Chinese] ... media because the protesters used only umbrellas and wet towels to protect themselves from the police's pepper spray and tear gas'. The Umbrella movement was a non-violent occupation that was organized by university students in Hong Kong, but came to include members from all parts of Hong Kong society. Despite the length of the occupation of public sites such as Tamar Park, Civic Square, Harcourt Road and Admiralty and the thousands of protesters involved, the Umbrella movement relinquished their occupation in late 2014. The Umbrella protests ultimately led to no political concessions from China regarding the Hong Kong electoral system. But, regardless of the outcome, we can use the Umbrella movement as an example of how smart objects can create phases that contribute to the dis-structive involution of the spaces, times and objects that these phases make partially intelligible.

Academic literature on the Umbrella movement debates the extent to which photos, video and text, captured by smartphones and communicated via social media played a role in shaping how members of the Umbrella movement organized themselves (Tsui 2015; Lee and Chan 2016). This work suggests that open, semi-public social networks such as Facebook and WhatsApp were important tools that enabled information about the sit-ins to spread among large groups of people, who specifically joined in the protests. As Lee and Ting (2015: 389) put it: 'Facebook ... [and] ... WhatsApp ... played an essential role in mobilizing the participants to join high-risk confrontations with the police'. But, at the same time, distinct from the centralized and semi-public platforms of Facebook and WhatsApp, the protesters also used more local means of communication, including Firechat and Telegram (Tsui 2015).

Firechat and Telegram are social messaging apps available for smartphones. Both apps utilize a wireless mesh network to deliver messages, with Telegram providing optional end-to-end encryption of these messages. Mesh networks allow communication between a series of devices using Bluetooth and wireless signals, but without being connected to a broader internet. In other words, you have to be physically present within the range of other phones in order to communicate and these communications cannot be accessed remotely. As Judah (2014: n.p.) summarizes, 'The Firechat app allows smartphone users to talk to one another "off-the-grid," in the absence of a mobile signal or access to the internet. By making use of Bluetooth and Wi-Fi, messages are spread in a daisy chain fashion, jumping from one user to the next. The system is particularly effective when large numbers of people are congregated together … .' The power of wireless mesh networks is that the more people use the app through their smartphones, the more powerful the network becomes. In Bland's (2014: n.p.) words, 'every new participant increases the network's range and strength. "Usually, the more people there are in the same location, the less connectivity you get," says Micha Benoliel, one of the app's creators. "But with our system, it's the opposite".'

The importance of Firechat to the Umbrella movement can be evidenced by reports of its usage. In Shadbolt's (2014: n.p.) words: 'In the first two weeks of the protests, between September 27 and October 10, the service registered 500,000 downloads in Hong Kong alone (61 per cent on Android and 39 per cent on iOS), 10.2 million chat sessions and 1.6 million chatrooms.' At certain moments of the protest over one hundred thousand accounts were being activated a day (Bland 2014), and on occasion 'thirty three thousand people in Hong Kong were using the app at the same time' (Judah 2014: n.p.).

Firechat creates a phase through the way that smartphones perturb one another, which discloses qualities that modulate relations of nearness and farness and then and now. The ability to send private messages and access public chat rooms is organized by the allopoietic structure of the Bluetooth and wireless antennas of the smartphones being used, meaning that messages can only be accessed when one phone successfully perturbs another. This creates relations of near and far organized around what the app refers to as 'Nearby'. In terms of metrical distance, nearby refers to approximately 61 metres, which is the average range of a smartphone's wireless antenna. When nearby, Firechat users can send messages privately between individuals or groups, or post publicly on chat rooms, which can accommodate up to ten thousand people at a time. But, in using Firechat there is no way of knowing when your phone is in range of another phone, because there is no way of knowing who around you has the app. For the human using Firechat, near and far thus become organized around a logic of spatial diffusion. The user can only move around and hope that their phone will successfully be perturbed by another, which will allow messages to be sent, but there is no intelligible boundary or marker that delineates the phase space of the app other than the connection icon within the app. At the same time, the app organizes relations of then and now around a modulating logic of dispersion and dilation. On the one hand, when Firechat is connected, time does not appear at all for the user as the message is seamlessly sent without delay. On the other hand, when Firechat cannot perturb another phone that has the app, time becomes spatialized as the number of perturbations that are required to send the message are unknown, which creates a sense of lag or delay for the user.

These phases and their logics of modulation shaped people's intelligibility of the Umbrella movement protests. For instance Eddie, who used the app during the protest, refers to how the phase constructed by the app altered where he went and when. For Eddie, relations of near and far became organized around the perturbations between phones, alongside the relationship between his body and the other humans who were part of the protest. For example, Eddie refers to 'bumping into' the phases of the Firechat groups. In his words 'When I got close to stands or tents where a speech is being given, I found people actually using Firechat to discuss the content and the speaker's points … [but at the same time you need to actually move around and] … "bump" into the right group to join a discussion' (Chao 2014: n.p.). For a protester named Daliguialt, the phase of Firechat encouraged a form of floating movement. As Daligault put it using Firechat was 'like being at a huge party … . You float from group to group and listen in. Eventually you find a group that is interesting. Or you can create your own' (Chao 2014: n.p.). The selective nature of the dissemination of messages, dependent on whether the user's phone could perturb another phone, served to draw bodies into spatial and temporal proximity with one another. Crucially these phases and the forms of movement they encouraged produced particular publics that were local to that situation. For Eddie and Daligault, the spatial and temporal relations disclosed by Firechat potentially drew them away and towards groups of people, until they found individuals who had common interests. Through the disclosure of this public, the phase potentially organized these individuals around a shared goal and in doing so aided the transformation of citizens or civilians into protesters.

Crucially, the production of this public was dependent on the phase that was specific to Firechat. The fact that the Firechat nearby feature only perturbed other phones within approximately 61 metres created a temporal instantaneity of response that enabled protesters to help one another and reinforce their sense of shared identity and resolve. As a protester who volunteered to monitor the supply stations at the Mong Kok protest site put it: 'We could meet up instantly to confirm who I was and what volunteers we wanted and to ensure they got the right information' (Chao 2014: n.p.). In one instance this enabled a resupply of ice to refrigerate food and water. 'We sent out a shortage message through Firechat and it did work .... As a result, a lot of people tried to contact me to see how to transit the ice packs' (Chao 2014: n.p.). In another case, the protester stated that 'the app helped to warn protestors about approaching police and g[ave] them time to take defense if need be' (Chao 2014: n.p.). More generally, other protesters suggested that Firechat helped enable a sense of a shared identity for the protest group through the way it could be used to 'inform other users where more umbrellas ... [were] ... needed for protection against pepper spray' attacks from the police (Mu 2014: n.p.). In doing so Firechat potentially worked to reinforce a sense of moral superiority for the protesters by distinguishing themselves as a well-organized and peaceful group, compared to what they considered to be a violent and over-reactive police force.

The phases of Firechat therefore contributed to a dis-structive involution by adding to the humans and non-human objects that sought to dismantle the spaces, times and objects that were made partially intelligible by the Firechat phase. In the case of the Umbrella movement, this dis-struction took place in a variety of ways. The objects that made up the streets, parks, boulevards and bridges where the protesters gathered in Tamar Park, Civic Square and Harcourt Road became dismantled through the reorganization of the street furniture and objects on the street. Streets such as Harcourt Road became sites of dwelling, where people sat and tents were erected. At the Admiralty site study areas were constructed using marquees with neon lighting powered by electric generators and pedal bikes, which allowed lectures and talks to be given and enabled younger protesters to study for exams and tests. In turn, relations of commerce and economic activity were dismantled as shops were forced to close for the duration of the protests and traffic rerouted around the protest sites.

The example of the Umbrella movement points to how dis-structive involutions aided by smart objects can lead to a variety of unanticipated consequences. For instance, taxi drivers and bus companies complained that the road blockages caused by the protests meant fewer fares (Davis 2014). But, with smaller numbers of vehicles on the road, air quality around the protest sites significantly improved. As The South China Morning Post (2014: n.p.) reported: 'Official data showed the health risk from air pollutants in the areas … [Causeway Bay, Central and Mong Kok] … was low rather than the typical high … [with a] … steep fall in levels of nitrogen oxide, one of the harmful pollutants emitted by diesel engines.' To summarize, the phases of Firechat and other mesh network messaging apps helped construct the specific space-times of the protests by altering the spatio-temporal intelligibility of various areas of Hong Kong for both the protesters and the general public.

Of course, I am not suggesting that apps such as Firechat caused the Umbrella protest. Nor am I implying that there is a 'one-dimensional cause–effect relationship between media and technologies' (Fuchs 2012: 384). What I am suggesting is that the use of Firechat as part of the Umbrella movement is a good example to think about how phases can generate dis-structive involutions of the space-times that these phases make at least partially intelligible. The dis-structive involution of Firechat enabled the protesters to reorganize objects in such a way to dismantle the dominant logics of the areas they were occupying in an attempt to generate a public around a particular political issue. Although the Umbrella movement did not ultimately achieve its goals, the dis-structive involution the protesters enabled did bring various issues around Hong Kong's relationship with China to the forefront of national and international consciousness.

Phase activism

This chapter has offered two techniques for responding to the phases constructed by smart objects. Both of these techniques could be termed forms of involution, in the sense that they attempt to alter phases by modifying the objects that contribute to the generation of these phases. As we have seen in the example of ride hailing services such as Uber, through structive involutions actors can alter or modify objects in order to change the kind of spaces and times that appear for users, which in turn can alter users' behaviour in those spaces. In the case of Uber, successfully modifying phases can have large financial repercussions. If London and other cities successfully regulate the rules around how space and time becomes intelligible in the app, this could have a direct impact on the number of people who can ride with and drive for Uber and thus impact Uber's profits. Indeed, an account of structive involution points to how important the spatio-temporal intelligibilities that phases disclose and modulate are for smart technology companies. But an account of structive involution also points to the fragility of these phases. Rather than banning Uber outright, Transport for London demonstrates how the smart objects that generate phases can be modified to radically alter the kind of spatio-temporal intelligibilities they disclose using quite simple and minor forms of regulation.

Developing dis-structive involutions, actors can use smart objects to dismantle and reorganize the spaces, times and things that smart objects make intelligible. In the case of the Umbrella movement, Firechat became a key object for the construction of spaces and times within which the objects and activities of protests were organized. The Umbrella protests are a key example of the dis-structive involution of a phase because Firechat worked to partially dismantle the space-times and objects that the circulation of messages on Firechat helped make intelligible. As people travelled around the protest sites, they 'bumped into' various phases of the app, which in some cases organized their activities to create a political public that dis-structed the commercial spaces of Hong Kong where the protests took place.

Based on these examples, the concept of involution points to a productive politics of phases and offers a blueprint for a kind of phase activism in relation to smart objects. Rather than limited to a critique that contextualizes and explains smart objects and their logics of spatial and temporal modulation within a history of neoliberalism and venture capitalism, involutions allow us to think productively about how phases can be actively challenged and altered in creative ways. Continuing this structive rather than critical logic of thought, the next chapter turns to discuss the ethics of smart objects. As smart objects gain the ability to perturb other objects without human intervention, this raises important questions about the limits of these perturbations and how smart objects should respond to situations in which they could cause harm.

7

Ethics

The ethical questions raised by the development, production and implementation of technical objects and systems are well known. As Manders-Huits (2011) usefully summarizes, engineers, designers, politicians and the general public regularly consider technical objects as neutral utilitarian things. In his words, 'when designing … technologies, the predominant, traditional focus of engineers is on functionality; the primary interests of engineers concern usability, efficiency, reliability, and affordability of (new) technologies. An engineer's principal concern is to make a technology which has the required functionality' (Manders-Huits 2011: 273). From a utilitarian perspective, an approach to technical ethics is a matter of ensuring that an object won't cause harm. As Verbeek (2008: 91–2) puts it, 'Engineering ethics and the ethics of design tend to … focus … on the importance of taking individual responsibility ("whistle blowing") to prevent technological disasters, and on methods that can be used to assess and balance the risks accompanying new technologies.'

From this perspective, technologies are understood 'in terms of their functionality: technologies are designed to do something, and if they fail to do so properly, they are badly designed' (Verbeek 2008: 92). However, while useful, such a utilitarian position 'fails to take into account … the impacts of such technologies on our moral decisions and actions, and on the quality of our lives' (Verbeek 2008: 92).

In other words, a purely utilitarian approach to technical ethics ignores the fact that all objects implicitly contain and transmit various values through their material structure and the types of practice they afford. Again, in Verbeek's words, 'Technology should not be considered to be value-neutral, but rather to have moral … impacts on humans and their environment.' Simple examples of the value-laden nature of technical objects would include objects that exclude or discriminate against particular groups of people, such as the public bridges built in Long Island, New York, that were so low that public transport could not pass under them, ensuring people from low-income backgrounds were unable to visit the parkways there (Winner 1980). To combat this kind of discriminatory design practice, Marshall (1999) argues that the ethics of technical objects requires us to ask questions about whose values are being represented in the object and how these values influence human behaviour and action. From this humanist value-based perspective, an ethical technical object would be one that, through its functioning, 'protect[s] and fulfil[s] the rights of individuals in a society' (Marshall 1999: 82).

Utilizing aspects of both a utilitarian and value-based approach to technical ethics, Introna (2007) posits an alternative post-human ethics of technology, which he terms a disclosive ethics (also see Brey 2000, 2005). A disclosive ethics recognizes the inherently folded nature of technical objects and human practices. Rather than considering technical objects as just creating social problems (such as accidents or disasters) or just reflecting social problems (such as class and income segregation), a disclosive ethics seeks to 'trace all the moral implications … from what seems to be simple pragmatic or technical decisions … through to social practices, and ultimately, to the production of particular social orders' (Introna 2007: 16). In other words, a disclosive ethics does not reduce the ethics of an object to the object alone, but seeks to understand the ethics that shape how the object is made and how that object goes on to shape the ethical behaviour of those who use it. From this position the human and technical cannot be separated. Disclosive ethics is less a matter of producing safe or unsafe objects (a utilitarian approach) or creating objects that do not discriminate (a value-based approach), than it is a matter of recognizing how technologies open and close different ways of entering into ethical relations with others. Introna gives the example of facial recognition algorithms used by security services to make this point. When searching a crowd, facial recognition algorithms can trigger false positives. As these algorithms find it easier to identify 'African-Americans, Asians, dark skinned persons and older people' this can result in these groups being 'subjected to disproportionate scrutiny' (Introna 2005: 84). Facial recognition technology thus encloses a specific group of people and at the same time implicitly reinforces pre-existing forms of bias. The ethics of facial recognition is not simply how it is used, but how its technical design groups people in order to shape the perception of the security agent using the technology, which can then create or reinforce existing modes of prejudice.

While understanding the ethics of technical objects in terms of utility, value and disclosure are important, this chapter suggests that taking smart objects seriously both requires and enables an alternative approach to ethics and technology. We could term this approach a phase ethics. Rather than focus on how objects are designed to ensure safety for their users, how they shape values or how they work to enclose human's relations with one another, a phase ethics concentrates on how smart objects identify and differentiate between entities in an environment. This ethics is not about just critiquing the design flaws in objects, the biased value systems of their creators or how technologies work to categorize groups of people. It is also about analysing and understanding the limitations of phases and sensitizing humans to these limitations in order to minimize perturbations that threaten the homeostasis of both humans and other objects.

In order to define and discuss phase ethics, the chapter examines a range of semi-autonomous and autonomous vehicle systems: specifically the autopilot system in Tesla vehicles and the Google smart car. The intentional and protentional structure of these semi-autonomous and autonomous vehicles complicate and confound existing academic debates around the ethics of technology. Specifically, ethical issues that were once considered to occupy the exclusive domain of the human can now be applied to smart objects such as autonomous vehicles as well. For instance, as smart objects become more complex and begin to respond to what is happening around them without human input, this raises a series of questions about how they make decisions that might inadvertently harm human and non-human others.

To examine these issues in detail the chapter focuses on autonomous vehicles through the thematic lens of the accident. Accidents provide important empirical examples to explore the ethics of smart objects because it is in moments of failure and breakdown that the intentional and protentional structure of smart objects and the assumptions behind their design appear most clearly (Latour 1996). In turn the concept of phase allows us to rethink ethics in relation to these accidents in ways that differ from either legal discourses of responsibility or the abstractions of ethical theory that are utilized in public reporting and discussion of these events. To expand and make sense of the notion of phase ethics, the rest of the chapter forms four sections. The section 'Ethics and Smart Vehicles' examines the ethical problems raised by autonomous vehicle systems and how they differ from the technical ethics discussed above. The section 'Phases and Accidents' discusses two examples of autonomous vehicle accidents in order to think through the objects involved in such accidents and how they produce phases. The section 'Phase Ethics' uses this analysis to define and illustrate the concept of phase ethics itself. The section 'Practicing Phase Ethics' offers constructive recommendations as to how a phase ethics can be practised in relation to the design and use of autonomous vehicles. To conclude, the chapter thinks beyond smart vehicles to consider how phase ethics might be understood in relation to smart objects more generally.

Ethics and smart vehicles

At its most simple, an autonomous vehicle can be defined as a vehicle that is 'designed to perform all safety-critical driving functions and monitor roadway conditions for an entire trip' (Reese 2016: n.p.). Examples of autonomous vehicles would be Google's self-driving car, which can complete journeys carrying passengers without human input. Alongside fully autonomous systems, the objects that support and enable autonomous navigation are also being utilized to enhance other types of vehicles, creating vehicles with varying degrees of semi-autonomy. Examples of vehicles with high levels of semi-autonomy would be Tesla's autopilot that is featured on its Model S, Model X and Model 3 cars. The autopilot has a high level of semi-autonomy because sensors on the vehicle, in combination with software, augment driver control and allow it to drive without direct human input. Examples of lower levels of semi-autonomy would be the Ford Active Park Assist system, which is available across a range of Ford models. The Active Park Assist has a low level of semi-autonomy because sensors around the car enable hands-free parallel parking, but this functionality cannot be utilized to engage in other forms of driving, such as riding on a highway.

In relation to the account of smart objects developed so far, autonomous vehicles and semi-autonomous systems, which from now on I will term smart vehicles, are perhaps the most advanced expression of the logic of smart objects currently available. This is because they have the most complex allopoietic and homeostatic structure and thus the widest capacity to perturb and be perturbed by objects compared to other smart devices. Like smartphones or smart security systems, smart vehicles utilize cameras and GPS, but can also incorporate radar, sonar, ultrasonic and laser sensors in tandem with on-board processors that enable them to respond to complex situations involving multiple objects 'via explicit pre-programmed instructions, a machine learning approach, or some combination of the two' (Goodall 2014: 101).

Using machine learning and simulations, smart vehicles are programmed with responses based on hundreds of thousands of hours of past use in an attempt to respond appropriately in their daily journeys (Michels et al. 2005). This combination of sensors and processing power gives smart vehicles complex forms of intentionality and protentionality and in turn the ability to produce phases that disclose intricate spatio-temporal relations for the vehicles and passengers riding in them.

Defined through the complexity of their intentional and protentional structure and the perturbations this structure affords, smart vehicles present moral and ethical problems that are difficult to account for within a humanist utilitarian or post-humanist disclosive perspective discussed in the introduction. This is because their ability to perturb and be perturbed by other objects that make up an environment provides them with the capacity to respond to situations without the explicit instruction or control of a human operator. This point is key, given the size, weight and speed these vehicles travel at, which makes it highly likely that an incorrect response could kill or damage human and non-human others. Of course, writers in the field of computing (Berleur and Brunnstein, 1996), robotics (Tamburrini 2009; Lin et al. 2011) and artificial intelligence (Anderson and Anderson 2011; Bostrom and Yudkowsky 2014) have been thinking through the ethical implications of smart objects that can respond in ways that might inadvertently cause harm for a long time. In relation to smart vehicles specifically, these questions are regularly framed around Foot's (1967) so-called trolley problem. The trolley problem is a thought experiment designed as a moral conundrum. According to Thomson (1985: 1395), the trolley problem begins with the premise:

> You are the driver of a trolley. The trolley rounds a bend, and there come into view ahead five track workmen who have been repairing the track. The track goes through a bit of a valley at that point, and the sides run … steep, so you must stop the trolley if you are to avoid running the five men down.

> You step on the brakes, but alas they don't work. Now you suddenly see a spur of track leading off to the right. You can turn the trolley onto it, and thus save the five men on the straight track ahead. Unfortunately ... there is one ... workman on the spur of the track. He can no more get off the track in time than the five can, so you will kill him if you turn the trolley onto him. Is it morally permissible to turn the trolley?

While the example was ultimately an experiment to test the outcomes and implications of different ethical systems in general (Edmonds 2013), the trolley problem has become a popular way of thinking through the literal implications of a smart vehicle crash or accident (for instance, see Hevelke and Nida-Rümelin 2015; Bonnefon et al. 2016; Lin 2016). Rather than a trolley and road workers, writers such as Nyholm and Smids (2016) suggest that smart vehicles responding to potential accident situations might have to engage in a decision similar to the driver of the trolley. For instance, if a car senses a pedestrian in the road, should it swerve to avoid the pedestrian, even if the swerve would mean the vehicle hits a wall and injures the passengers? This problem becomes more complex when more variables are included. As Markoff (2016: n.p.) wonders, 'Is it acceptable for an AV ... [Autonomous Vehicle] ... to avoid a motorcycle by swerving into a wall, considering that the probability of survival is greater for the passenger of the AV than for the rider of the motorcycle? Should AVs account for the ages of passengers and pedestrians?'

Within the literature on smart vehicles such ethical questions are often presented through a distinction between a deontological and consequentialist approach to ethics (Goodall 2014). A deontological account revolves around the idea that any ethics must be based on rules that guide what people do and that an action would be ethical as long as it was based on sound principles, regardless of the outcome (Crisp 1995). In the case of a smart vehicle, the ethical system programmed into the vehicle might be based on the notion that it has to protect its driver at all costs. If the vehicle were presented with a situation in which crashing into a wall at high speed would avoid hitting two pedestrians, then the ethical decision would be to protect the driver by hitting the pedestrians, even though it could result in the injury or death of the pedestrians. Distinct from a deontological position, a consequentialist account is based on 'the central principle … [of] … the greatest good for the greatest number' (Barbour 2011: 114). A consequentialist perspective would suggest that it is the consequence of a rule, rather than the rule itself that makes an action ethical or otherwise. Returning to the same example, from a consequentialist perspective, if crashing into the wall would save the life of two pedestrians but kill the driver, then the vehicle choosing to crash into the wall would be considered the ethical decision.

In relation to smart vehicles, these moral and ethical questions are quickly folded into legal understandings of fault, responsibility and liability. For instance, 'If a manufacturer offers different versions of its moral algorithm [perhaps allowing the owner to set the vehicle to hit a pedestrian in order to minimise harm to the passengers in the event of a potential crash], and a buyer knowingly chose one of them, is the buyer to blame for the harmful consequences of the algorithm's decisions?' (Markoff 2016: n.p.). On the basis of such a quandary, a number of vehicle manufacturers have already made public statements as to what their smart vehicles will prioritize in the event of an unavoidable crash or accident. For example, Mercedes have claimed that any of their self-driving cars will seek to protect the passengers inside the car over pedestrians or bystanders in the event of an accident, regardless of the fault of the driver (Markoff 2016).

What is notable about various theoretical and legal approaches to the ethics of smart vehicles is that while different in many ways, they all operate within a broader utilitarian ethical framework. From this position, the ethical questions around smart vehicles ultimately focus on the potential damage they can cause and how this damage might be minimized through processes of design and testing. However, if we analyse the ethics of smart vehicles through actual real-world accidents, using the concept of phases, a rather different picture emerges. The ethics of smart objects 'cannot be entirely reduced to the intentions behind their design and use' (Verbeek 2008: 91) because of the inherent contingencies involved in the phases smart objects produce, which depend on how a smart object perturbs and is perturbed by things outside of the controlled zones of thought experiments, laboratory tests and simulations (Hutchins 1996).

Phases and accidents

According to Campbell (1997: 31), 'accidents … are interventions – slippages – in the control we attempt to exert over the course of lives'. In relation to technical objects, Virilio suggests that every technology brings with it the unique potential for a new kind of accident. In his words, 'the shipwreck is … the … invention of the ship, and the air crash the invention of the supersonic airliner, just as the Chernobyl meltdown is the invention of the nuclear power station' (Virilio 2007: 5). From this position, the particular components of objects and how they are assembled together also create unique conditions for failure. In the case of Chernobyl, the reactor that contained the graphite rods that are used to generate nuclear energy also acted as a potent source of potential explosion and radioactive fallout. Accidents are therefore useful to understand how technical objects can respond in ways that can't be anticipated in their design and testing. In relation to smart vehicles in particular, accidents provide the context for understanding processes of recurrent causality, where the phases that smart vehicles generate in order to travel can also work to destroy these phases and the objects that created them. The following section explores two accidents in particular, a fatal crash involving a Tesla Model S and a more minor crash involving a Google car. Through these examples it will become clear that to develop an ethics sensitive to smart vehicles requires that we understand the phases the vehicles produced in the particular situation of the accident, rather than focus on what the vehicles were programmed or designed to do in advance of these situations.

a) Tesla Model S accident

On 7 May 2016, a Tesla Model S travelling in Florida in the United States crashed, killing the driver, Joshua Brown, upon impact. This was the first time a Tesla Model S was involved in a deadly accident. The car was being driven using the Tesla's autopilot mode. The autopilot mode offers drivers what Tesla terms a 'semi-autonomous' driving system. This means that if the road conditions are correct the car can autosteer, change lane, manage its speed and park without the direct input of a driver.

The crash took place on the US-27A highway in Florida and involved a trailer engaging in a left-hand turn onto NE 140th court in front of the Tesla, which seemingly failed to spot the trailer and drove at full speed into it. With autopilot activated, the car's roof was crushed and the vehicle continued to travel, veering off the road striking two fences and coming to a standstill after hitting a power pole (Golson 2017). Reports of the crash centred on assigning responsibility to either the driver or the autopilot system and fed into an emerging narrative about users overly relying on the autopilot system to drive for them. Despite visual safety warnings presented on the car's heads-up display during the autopilot mode, which reminds drivers to 'always keep your hands on the wheel' and 'be prepared to take over at any time' (Oremus 2016: n.p.), videos have emerged showing drivers riding hands free and even asleep at the wheel with the autopilot mode activated (Smith 2016).

A utilitarian or de-ontological approach to the crash might focus on why the vehicle did not make the correct decision and alert the driver as to the presence of the trailer in the middle of the road. However, rather than assuming that the autopilot mode had all the correct information it needed to make a decision that would minimize perturbations to the car and driver, the concept of phase asks us to step back to examine the conditions of possibility of the crash. Specifically we need to understand the intentional and protentional structure of the Tesla and analyse the sensor components that make up the autopilot system. In turn, we can then focus on which qualities of objects the perturbations from the sensors disclosed and how these qualities shaped what was intelligible for the vehicle and driver.

In the Model S Joshua Brown was driving, at least four components can be identified that are linked to the autopilot system: the radar, the ultrasonic sensors, the optical cameras and the GPS system, all of which are used to detect the presence and location of objects.[1] The specific way each of these components perturb and are perturbed by other things disclose different qualities of objects, which shape the intelligibility of space and time for the Model S. Specifically, the radar perturbs and is perturbed by other objects through emitting and receiving radio waves via an antenna. The radar system is complimented by a 360° ultrasonic sonar. Sonar operates by emitting very high-frequency sound waves via a speaker that perturbs objects. Depending on the qualities of the surfaces it encounters, these sound waves then disclose different qualities of the surface, which perturb a microphone in the ultrasonic sensor. Dependent on the time it takes to return, the perturbations from the radar and sonar can be used to determine the location, size and speed of an object. In addition to radar and sonar, optical cameras on the car are perturbed by light that enters their lenses, which then perturbs image recognition software to determine which patterns of perturbation are objects. Finally all these sensors work alongside GPS tracking to enable the car to recognize turns and junctions in the road according to pre-written digital maps.

---

[1] In later revisions of the Model S, the radar system is the primary control sensor and uses software to construct a 3-D image of the world, rather than rely upon confirmation from the camera.

Remembering that each of the Tesla's sensors can only disclose particular qualities of the objects they perturb and are perturbed by, we then have to take into account the other objects involved in the crash. For instance, think of the high-sided trailer and the sky at the time of the crash. Tesla chief executive Elon Musk suggested that the crash may have been caused by the autopilot system not recognizing the trailer as an obstacle. As Fleming (2016: n.p.) writes, Musk suggested that the autopilot system's camera may have been unable to 'isolate the image of the trailer from the bright sky behind it', which meant the Tesla failed to recognize the trailer and did not apply the brakes. From a phase perspective, the bright light is itself an object produced by the sun, which is enabled by a sky relatively clear of clouds or other atmospheric phenomena that would block or occlude the sunlight from perturbing the vehicle's camera sensor. If we follow Musk's initial suggestion as to the cause of the crash, the inability of the autopilot system to perturb the trailer and sky and produce qualities that would identify them as distinct objects meant that the Model S constructed a phase between objects in the environment, while missing a crucial one (the trailer of the truck). In turn, this phase meant that the vehicle was unable to assign any references (such as danger) to the trailer and so did not react to the presence of the trailer. The failure of the autopilot may have been the incapacity of the camera to perturb the sky and trailer in a way that would disclose qualities that would enable the camera to differentiate between them and thus recognize the presence of two distinct entities. The phase constructed between the Tesla and objects in the broader environment thus created a very specific spatio-temporality. Space became intelligible for the Tesla as an opening in front of the camera sensor. Time, understood as an appearance shaped by the changing qualities of objects, emerged through a relation between the current speed the car was travelling at (74 kilometres per hour), the estimated braking distance it would take to stop while travelling at this speed and the camera system's assumption that there was an opening in front of the car, which meant the braking system did not activate.

Analysed as a phase, the accident was not a spatial or temporal event in which something happened, but an aligning of objects to construct a phase in such a way that failed to disclose important qualities of objects that were necessary for the continuing homeostasis of the driver and vehicle. If the driver had been travelling without autopilot then they may have assigned references to the trailer and presumably applied the Tesla's brakes to allow the truck to cross the highway. While the autopilot system created a phase that generated spatio-temporal relations of near and far, then and now, which primed what was intelligible for the vehicle, the driver of the vehicle still had to ultimately assign references to the space-times the phase disclosed. In this case, a failure of the driver to assign these references ultimately resulted in their unfortunate death. In our next example we will see that smart vehicle accidents are not always linked to the way the phases smart objects generate differentiate between objects, but can also occur when the intentional and protentional structure of the vehicle cannot deal with the contingency of other drivers.

b) Google car accident

Google's self-driving car programme began in 2009. The Google car is a modified road vehicle that boasts a set of sensors that enable it to drive passengers from point A to point B without any human assistance or intervention. To do this, the car uses Light Imaging Detection And Ranging (LIDAR) as well as radar and sonar systems. LIDAR, according to Whitwam (2014: n.p.), 'bounces a beam off surfaces and measures the reflection to determine distance. The device used by Google – a Velodyne 64-beam laser – can also rotate 360-degrees and take up to 1.3 million readings per second … and is highly accurate up to a range of 100 meters.'

In concert with sonar, radar, GPS and data analysis, the LIDAR system allows Google (2017: n.p.) to boast that their smart vehicles have travelled over two million miles on public roads in the United States, with the aim 'to make our roads safer, free up people's time, and improve mobility for everyone'. However, despite Google's claims as to the safety of its smart driving, on 14th February 2016, a Google vehicle was involved in an accident that Google admits was partially the vehicle's fault. An extract from the Department of Motor Vehicles accident report gives details of the incident. Although lengthy, it is worth reproducing in full here:

> A Google Lexus-model smart vehicle ('Google AV') was traveling in smart mode eastbound on El Camino Real in Mountain View in the far right-hand lane approaching the Castro St. intersection. As the Google AV approached the intersection, it signalled its intent to make a right turn on red onto Castro St. The Google AV then moved to the right-hand side of the lane to pass traffic in the same lane that was stopped at the intersection and proceeding straight. However, the Google AV had to come to a stop and go around sandbags positioned around a storm drain that were blocking its path. When the light turned green, traffic in the lane continued past the Google AV. After a few cars had passed, the Google AV began to proceed back into the center of the lane to pass the sand bags. A public transit bus was approaching from behind. The Google AV test driver saw the bus approaching in the left side mirror but believed the bus would stop or slow to allow the Google AV to continue. Approximately three seconds later, as the Google AV was re-entering the center of the lane it made contact with the side of the bus. The Google AV was operating in smart mode and traveling at less than 2 mph, and the bus was travelling at about 15 mph at the time of contact.
>
> The Google AV sustained body damage to the left front fender, the left front wheel and one of its driver's-side sensors. There were no injuries reported at the scene. (Lirmson 2016: 2)

The accident seems to have been caused by the phase the Google vehicle disclosed, which in turn incorrectly guided its response to other objects that surrounded it. In terms of phase time, the accident was perhaps caused by the inability of the car to modulate between intentions and protentions that would allow it to recognize that the bus was likely to turn. In the case of the Google car, the protentions of the vehicle would be determined by the phase memory of all the recorded previous journeys Google vehicles had gone on, which shaped its machine learning algorithms and informed the rules it used to navigate. The Google car's intentions would have been based on the qualities disclosed by the perturbations from its LIDAR sensors in concert with its machine learning algorithms, which it could compare to its phase memory in order to respond appropriately.

The phase time disclosed by the phase memory and protentions of the vehicle were not accidental, but the direct result of the vehicle's protentionality being altered a few weeks prior to the crash. In Google's (2016: 1–2) words,

> Several weeks ago we began giving the self-driving car the capabilities it needs to do what human drivers do: hug the rightmost side of the lane. This is the social norm because a turning vehicle often has to pause and wait for pedestrians; hugging the curb allows other drivers to continue on their way by passing on the left. It's vital for us to develop advanced skills that respect not just the letter of the traffic code but the spirit of the road.

In this case, the attempt by Google to respond to the social norms of driving generated a temporal intelligibility of the vehicle that also shaped the spatio-temporal intelligibility of the driver of the bus as well. For the Google car, the perturbations from the LIDAR sensor disclosed that there was a distance between itself and the bus that would allow the car to continue on its journey. At the same time, recognizing that the vehicle was being courteous by pulling slightly to the left of the lane, the bus driver may have assumed that this courtesy would include allowing his bus to pass. In Google's words, while the LIDAR sensor had been perturbed by the bus and recognized it as a bus, the car 'predicted that … [the bus] … would yield to us because we were ahead of it' (Google 2016: 1–2).

This modulation between intention and protention shaped the spatio-temporal intelligibility of the phase for the car and bus driver. In the phase of the accident, space appeared through the anticipation of the driver and protention of the car, who both assumed that a gap would be open to move into. This gap was partially disclosed in advance, via the way time appeared through the perturbations of the LIDAR sensor, which was perturbed by the presence of the bus and its speed of travel, but assigned the incorrect references to the bus (that it would stop or give way).

Phase ethics

The two events described above serve as landmarks in the emerging history of smart vehicles. But for our purpose they also provide important examples for thinking through the ethics of smart objects. Analysing these accidents in terms of phases has demonstrated three issues that call into question the suitability of utilitarian and deontological accounts for understanding the ethics of smart vehicles. First, both the Tesla Model S and Google car crash have demonstrated that discussing ethics using a language of 'decision' is not particularly helpful. A smart object does not organize its action through a unification of the data it receives into a coherent picture or image of the world to make a decision. As we have seen these vehicles are not intelligent, they simply utilize sensors that have the capacity to perturb and be perturbed by other things in a range of different ways.

Second, theorizing ethics as a decision based upon a spatial and temporal moment or event that runs according to a predetermined set of rules is not helpful either. Rather than taking place in space and time, these accidents were potentially caused by the way the Tesla Model S and Google car made space and time intelligible, both for the vehicles themselves and the drivers of these vehicles. In other words, one way to analyse these accidents is to examine how the smart vehicles differentiated or failed to differentiate between objects. In turn these differentiations disclosed space-times in which both the vehicles' and drivers' homeostasis were brought into question.

Third, the concept of phases undoes notions of distinct objects that underlie utilitarian and deontological notions of ethics. Smart vehicles do not have a strict material boundary that separates them from the broader environment. Phases extend and intensify through the perturbations of vehicle sensors and these vehicles can only travel and move through the way they perturb and are perturbed by other objects in the environment, such as road signs, pavements, kerb stones, road markings, light conditions, other traffic, other drivers' interpretation of the highway code and so on. From this perspective there is no single point or moment of encounter where the genesis of an accident can be located. Instead we have to trace how different objects perturb one another in order to identify the qualities they disclose.

Rather than a failure of decision-making in relation to the pre-written ethical framework of the vehicle that forms part of either its machine learning programming or its sensors, these accidents can perhaps be understood as the outcome of the constitutive contingency of the way the Tesla and Google car perturbed and were perturbed by qualities of objects that made up the environment they were travelling in. A deontological or utilitarian ethics of these vehicles cannot take account of this contingency, precisely because these models assume that smart vehicles involved in accidents have the necessary information to make an informed decision and will make this decision according to such information. Exposing the limitations of either a deontological or consequentialist analysis of smart vehicle accidents through a language of phases points towards an alternative ethics of smart vehicles, which we can term phase ethics.

Phase ethics recognizes the impossibility of creating smart objects that can accurately apply distinctions between good and bad or right and wrong according to an ethical set of rules or code concocted in advance and activated via the phase memory of an objects' machine learning. Instead, phase ethics focuses on how smart objects sense other objects. A phase ethics asks us to analyse what qualities of other objects smart objects disclose and highlights the always limited nature of this disclosure. Practically, a phase ethics can aid the development of objects and practices that minimize perturbations that challenge or disrupt other humans' and non-humans' homeostasis. This is not just a question of design, value or disclosure, but of how smart objects make space and time intelligible for both themselves and humans. What does this account of phase ethics mean in practice? To answer this question let's suggest some practical recommendations such an ethics would offer to debates around smart vehicles based on issues of liability, environments and language.

Practising phase ethics

a) Liability

Practising a phase ethics would require a rethinking of current notions of liability in relation to smart vehicles. In the United States, legal discussions of liability are based around a definition of an accident based on the 'no-reasonable person' standard. When an individual driver is clearly not at fault and could not have avoided an incident, it is considered an accident. But, in emerging discussions of smart vehicles no such liability conditions are being questioned. Goodall (2016: n.p.) explains this in the following way:

> Today no court ever asks why a driver does anything in particular in the critical moments before a crash. The question is moot as to liability – the driver panicked, he wasn't thinking, he acted on instinct. But when robots are doing the driving, 'Why?' becomes a valid question. Human ethical standards, imperfectly codified in law, make all kinds of assumptions that engineers have not yet dared to make. The most important such assumption

> is that a person of good judgment will know when to disregard the letter of the law in order to honor the spirit of the law.

Goodall suggests that what is required for smart vehicles to be ethical is that they are 'taught' good judgement in order to make better decisions. In his words, 'What engineers must now do is teach the elements of good judgment to cars and other self-guided machines – that is, to robots' (Goodall 2016: n.p.). A phase ethics would disagree with such a position. Deontological and consequentialist discussions of the ethics of smart vehicles are based on a fallacy that smart vehicles are somehow all seeing and knowing and can make the right decision if only they are programmed with the correct moral code. This is reflected in work by Lin (2013: n.p.) who argues, 'Human drivers may be forgiven for making an instinctive but nonetheless bad split-second decision, such as swerving into incoming traffic rather than the other way into a field. But programmers and designers of automated cars don't have that luxury, since they do have the time to get it right and therefore bear more responsibility for bad outcomes.'

A phase analysis of the Tesla Model S and Google car crashes shows this is not the case. When Goodall suggests that 'vehicle[s] must at times make ethical choices smartly, either via explicit pre-programmed instructions, a machine learning approach, or some combination of the two' (Goodall 2014: 101), he is already presuming that this choice will be based on a correct or total amount of information that the vehicle will use to make the said choice. But, the phase spaces and times smart vehicles disclose are just as limited and fallible as human sensory perception. As the discussion of the Google car crash and our earlier theorization of protentionality in Chapter 2 demonstrates, smart objects do not anticipate, but rather await qualities disclosed by the perturbations from their sensors and these sensors only offer a partial disclosure of an environment. No matter how many hours of test-driving or simulation data a designer uses to programme a vehicle, the designer can never anticipate the contingency of the qualities of objects that are disclosed by smart vehicles sensing components. As I have argued throughout the book, this is because objects always have the potential to be perturbed in a way not envisaged by their designers and thus can disclose qualities (and in turn spatio-temporal relations) that exceed a smart object's ability to sense them.

A phase account asks us to move beyond an ethics based on a moment of decision-making in light of correct or absolute information about a situation to focus on an object's capacity to sense. Just as there is a no-reasonable person argument in US law regarding human liability, we might suggest that there can also be a no-reasonable robot argument regarding smart vehicle liability as well. A smart vehicle can never be reasonable, regardless of the complexity of its intentionality, protentionality or phase memory because the phase the vehicle generates will always be partial and contingent. The best a smart vehicle can do is be transparent in how it attempts to avoid perturbations that reduce an object's capacity for homeostasis, based on the limitation of the spaces and times it makes intelligible. In other words, the phase of a smart vehicle cannot be assumed to be more accurate or less partial than a human's sense perception. Rather than a matter of immediately assigning legal blame or responsibility to the smart vehicle itself, a phase ethics suggests that accidents should be understood through the relationship between the phases smart objects generate – how these phases make space and time intelligible for the vehicle and how this intelligibility is communicated to the driver. Only once these factors have been identified can liability be fairly apportioned.

b) Environments

A second recommendation of a phase ethics of fully smart vehicles would suggest the need to develop environments that are designed to work with the limitations of the phases that smart vehicles produce. As Verbeek (2008: 101) suggests, 'When artifacts have moral relevance and embody a specific form of moral agency, ethics cannot only occupy itself with developing conceptual frame-works for moral reflection, but should also engage in the development of the material environments that helps to form moral action and decision-making.' In the case of the Google car, the 'material environment' that Verbeek highlights would involve the road system on which the car drives, including clear lane markings present on every street, uniform heights and colours of road signs and potentially an integrated system that could be used to monitor the relations between vehicles as they move around the road. While an environment can never stop the potential for a crash to occur, recognizing that ethics is a massively distributed negotiation between objects, rather than a decision-making process contained within a single object would be a helpful move.

In relation to vehicles with semi-smart functionality, such as the Tesla Model S, a phase ethics would suggest that the cockpit of the vehicle as well as the road system is key to 'form[ing] … moral action and decision-making' (Verbeek 2008: 101). Specifically the cockpit could be designed to continually perturb the user and help them assign references to the spaces and times disclosed by a phase in order to give them the best chance to respond in the event of a potential accident. The beginnings of such logic can be seen in Tesla's updated firmware for its autopilot system. Where the original version of the autopilot system gave text warnings to the driver reminding them to stay in control of the vehicle when using autopilot mode, the current firmware (v8.0) will slow down and stop if it senses that the driver has taken their hands off the wheel. In turn, if the driver continues to 'repeatedly ignore Autopilot prompts, autosteer will disengage and remain unavailable for the duration of the drive' (Gene 2017: n.p.).

c) Language

Third, a phase ethics would caution against over-emphasizing the capacities of smart vehicles, which gives drivers a sense that smart vehicles have a kind of agency or intelligence that they, in fact, do not have. Unfortunately it would appear that some manufacturers are at least implying such capacities in the marketing and advertising of their products. For instance, Tesla are keen to emphasize that their vehicles' sensor components are superior to human sensory capacities. In their words, the autopilot system 'provides a view of the world that a driver alone cannot access, seeing in every direction simultaneously, and on wavelengths that go far beyond the human senses' (Tesla 2017: n.p.). Tesla reiterate this point through the way they visually represent and measure the capacities of the Tesla Model S's sensors in the form of metrical distance in their marketing material. A diagram on Tesla's website shows the Model S as a small black dot in the middle of the screen, with a range of coloured cones protruding from the car. The narrow forward camera is represented as a light blue cone with a label stating 'Max distance 250m', the radar is a light grey cone with a label stating 'Max distance 160m', the forward-looking side cameras are a light green cone with a label stating 'Max distance 80m' and so on. This simple diagram emphasizes the extensive power of the sensors while minimizing, via the size of the car on screen, the capacities of the driver who is presumably smaller than the black cylinder that represents the car.

Despite Tesla's marketing material, a phase analysis has demonstrated that it is not helpful to understand smart objects' sensory capacities in the form of metrical distance. The phases that smart objects generate are not about sensing space and time as a set of metrics or amounts or about gathering data together to form a coherent picture of the world in a process akin to human consciousness. Rather the phases of smart objects are defined by their capacity to differentiate between objects and assign the correct references to those objects in order to make distance sensible and intelligible. As the Tesla Model S accident shows, it does not matter how 'far' a sensor can reach, if that sensor cannot differentiate between objects in order to produce a phase and so enable a car or driver to assign the correct references to those objects.

In a similar vein, a phase ethics would suggest that the limitations of the phases disclosed by a vehicle cannot simply be overcome by the development of more advanced or powerful sensors. For instance, in its latest revision of the Model S, Tesla (2016b) have stated that they will no longer use the visual camera to confirm the presence of objects. Instead the radar system is positioned as key to object detection. The length and frequency of radar waves means that radar can 'see through' a variety of what humans consider to be solid objects. As Tesla (2016b: n.p.) suggest, this is useful as it allows the autopilot system to sense visually obscured objects and allows the car to 'travel easily through fog, dust, rain and snow'. However, the radar itself is not a panacea. As Tesla themselves recognize, while a radar can see through fog, when encountering metal objects that are dish shaped, this can

> amplif[y] … the reflected signal to many times its actual size. A discarded soda can on the road, with its concave bottom facing towards you can appear to be a large and dangerous obstacle, but you would definitely not want to slam on the brakes to avoid it. Therefore, the big problem in using radar to stop the car is avoiding false alarms. Slamming on the brakes is critical if you are about to hit something large and solid, but not if you are merely about to run over a soda can. Having lots of unnecessary braking events would at best be very annoying and at worst cause injury. (Tesla 2016b: n.p.)

While Tesla suggest they use a range of different sensors to overcome this issue, their statement reminds us that smart object sensors can only ever disclose particular qualities of other objects, rather than access the totality of objects themselves. Smart vehicle manufacturers would be wise to remember this and avoid making claims that insinuate that their vehicles have greater or more extensive sensory capacities than their human drivers.

Beyond over-emphasizing smart vehicles, sensing capacities, a phase ethics would suggest that companies must also be very careful about how they name the systems that control their smart vehicles. Critics such as Taub (2016) have pointed out the name of Tesla's autopilot system is misleading and implies that the driver does not need to be in direct control of the vehicle. In his words, 'Drivers who think they now own a self-driving vehicle are sadly, and perhaps dangerously, mistaken' (Taub 2016: n.p.). Tesla (2016a: n.p.) argue that users recognize that the autopilot system is not self-driving by pointing to a survey conducted with German Tesla owners that demonstrated '98% of customers … understand that when using Autopilot, the driver is expected to maintain control of the vehicle at all times'. Despite Tesla's assertions, demo videos available on the Tesla website clearly advertise self-driving as a key feature of the vehicle. A video embedded on the site's autopilot page begins with the words 'the person in the driver's seat is only there for legal reasons. He is not doing anything. The car is driving itself' (Tesla 2017). The video shows a beginning to end the journey of a Tesla vehicle with the driver seated with his hands by his side and his feet away from the pedals. Such videos would seem to encourage drivers to put far too much faith in the phases smart objects generate. A phase ethics of smart objects would recognize the dangers of such faith and instead insist on describing and marketing these technologies using a language of partial sensing rather than autonomy or automation. This would be one simple way of highlighting the fragility of smart vehicles and the phases they produce.

Phase ethic futures

The complex intentions, protentions and phase memories of smart objects such as smart vehicles require the development of new modes of ethical analysis. In this chapter, I have proposed one such mode, which I have termed a 'phase ethics'. This phase ethics is distinct from either utilitarian, disclosive, deontological or consequentialist notions of ethics that have been used to discuss technology in general and smart vehicles specifically. While different in many ways, what unites either a utilitarian, disclosive, deontological or consequentialist account of the ethics of smart vehicles is the assumption that the vehicle itself has equal or better access to the necessary information to make an informed ethical decision regarding what action to take in the event of a potential accident than a human driver would have. A phase ethics asks us to step back from questions of utility, consequence or moral absolutes. To understand what is ethical requires we begin by understanding the type of phase an object can produce and recognizing the inherent limitations of this phase.

Work in computing and artificial intelligence often assumes that while human action is fallible, machine action can be honed to produce the least damaging outcome (Tavani 2011). In legal interpretations of vehicle accidents, empathy and reasonable doubt is given to the human driver but not the vehicle itself. A phase ethics would suggest that it is necessary to give reasonable doubt to the machine as well, but only if we define reasonable doubt as a clearly articulated recognition by manufacturers of the limitations of the phases smart vehicle can generate, which drivers and other road users understand. While the name implies otherwise, autonomous vehicles are not really autonomous. Smart vehicles are composed of multiple components such as sensors and cameras as well as machine learning software and can only intend and protend what is happening through the qualities disclosed by the way their sensors perturb and are perturbed by other objects. It is only when we make the error of assuming that smart vehicles are autonomous that ethical questions around decision-making become collapsed into simple dichotomies of choice around what to hit and what to miss.

A phase ethics of smart objects will only become more necessary and pressing as the intentional and protentional structure of smart vehicles becomes increasingly complex and their presence in the environment more ubiquitous. For instance, think of the development of Amazon Prime Air. Amazon Prime Air is an autonomous delivery system that uses drones to deliver packages up to five pounds in weight in less than thirty minutes. Although in its initial trial stages, Amazon hopes to offer Prime Air to millions of customers, potentially creating a situation in which thousands of drones are flying simultaneously at any one time (on drones more broadly, see Jackman 2016; Shaw 2016). The complexities of flight offer additional issues and dangers compared to road-based autonomous vehicles. For instance, the potential for crashes are not limited to roads or highways and could potentially happen anywhere, causing damage to both human and non-human others. Furthermore, minor failures in various components of drones result in the potential for more damaging accidents. Rather than coming to a standstill, as in the case of a car, a drone would drop from the sky and potentially hit other objects with little warning. In this case a phase ethics would be required to think through the kinds of phases drones generate and how they differentiate between objects in order to respond to situations.

The previous seven chapters have defined smart objects, demonstrated how they construct phase spaces and times, analysed how they disclose issues as political, and illustrated how individuals and groups can respond to these issues in terms of politics and ethics. In the following final chapter, I reflect on the arguments presented across the book and suggest that the concept of phases is helpful for rethinking taken-for-granted accounts of technical networks beyond the specific example of smart objects.

8

After Networks

This book has argued that smart objects, understood as intentional and protentional devices, produce phases. These phases can be defined as space-times disclosed by the way smart objects perturb and are perturbed by other things, which modulate the spatio-temporal intelligibility of the humans and non-humans who live alongside smart objects. Companies involved in the production of smart objects aim to create phases that modulate the spatial and temporal intelligibility of humans to generate consumer demand and regulate both consumer and producer behaviour. Companies work to achieve this goal by utilizing a series of logics, through which space and time are diffused, partitioned, enveloped, gradated, dispersed and dilated in different ways for different purposes. As I examined in relation to ride-sharing apps such as Uber, space and time are enveloped and gradated to disclose relations of spatial and temporal proximity between Uber drivers and Uber customers to create the appearance that their service is the fastest and most convenient to use. In relation to smart objects such as the Dyson 360 Eye vacuum cleaner, space and time is diffused and dispersed in order to attempt to minimize the intelligibility of the cleaner so it appears efficient and untroublesome.

Despite the flexibility of these multiple techniques, the book has suggested that phases cannot and should not be reduced to the constructive logic of smart object manufacturers and designers. Instead I have argued that phases can be helpfully understood, following Nancy and Barrau (2015), through the logic of struction. Opposed to the instrumental, goal-orientated logic of construction, struction understands smart objects as a mass or heap of entities that do not neatly connect to form a holistic network of references. Focusing on the structive nature of phases asks us to examine how phases can emerge in all kinds of unanticipated ways and create political publics that exceed and confound the most powerful or wealthy hardware manufacturers and app developers. As such, the kinds of political publics that emerge through phases can be both intentional, in the sense that the creators of the smart objects that generate these phases have a particular aim in mind, but they can also be contingent or accidental. In the case of the Wishaw telephone mast, discussed in Chapter 5, the mast itself worked to create a phase and generate a series of references for the residents that the owners of the mast did not want to be made. In the case of Hong Kong's Umbrella movement, discussed in Chapter 6, the Firechat app contributed to the involution of a phase, in the sense that Firechat enabled the conditions for the dismantling of the very spaces, times and objects it served to partially disclose. In these cases, the logics of diffusion, partition, envelopment, gradation, dispersion and dilation utilized by the creators of smart objects to modulate spatio-temporal intelligibility can be drawn upon to develop alternative phases and new political publics and ways of life.

As well as demonstrating how phases can be understood through a constructive and structive logic, the concept of the phase also has broader value beyond a discussion of smart objects per se. Specifically the concept of phases provides an alternative to two wider interlinked accounts of networks. On the one hand, the term network has become part of a general conceptual imaginary to understand relations between objects in general. On the other hand, the term network is used to understand media as a form of infrastructure or set of communication protocols between technical objects. With these two notions of network in mind, the section 'Networks and Phases' of this final chapter outlines and summarizes key distinctions between the concept of the phase and the concept of networks and how the notion of phases provides an alternative to both general social theories of networks and specific technical understandings of the term beyond the specific example of smart objects. The final section offers some brief closing remarks.

Networks and phases

On a discursive level, the term phase is a useful compliment to the term network. While so much digital infrastructure is still very much wired, the term phase more easily evokes the shift towards wireless, ambient, atmospheric and diffused modes of contemporary media communication that the term network, with its connotations of stasis and fixity misses. But, perhaps more importantly, the term phase also offers an alternative narrative of technical objects that is distinct from accounts of networks based upon a logic of points and lines or nodes and edges. There are at least three key differences between the concepts of networks and phases. First, while networks are often considered as material constructions, phases are sets of objects. Second, while networks are defined by the presence of objects, phases are defined by the accessibility of objects. Third, a network's capacity to change is defined by shifts that occur either internal or external to a system, while phases' capacities to change are defined by recurrent causalities that emerge from the interiority of objects, rather than some process outside of an object. Examining each of these differences in detail provides further points of clarification regarding the arguments developed across the book and also offers useful axioms for others who are interested in using the concept of phase in their own work.

a) Networks are material constructions. Phases are sets of objects.

Returning to the climatological impulse of Galloway and Thacker (2007) that opened the book, a theory of phases disrupts understandings of networks as material assemblages. Beyond media studies, understanding the world as a set of material assemblages has become widespread. As I suggested in the introduction, this new materialism focuses on materials as relational, structured by events of contact and connection. Anderson and Wiley (2009: 318) sum up this position when they explain that a turn towards new materialism involves

> an attunement to how heterogeneous materialities actuate or emerge from within the assembling of multiple, differential, relations and how the properties and/or capacities of materialities thereafter become effects of that assembling. … Materialities as apparently varied as a butterfly … or an ideal … come into being from within the event of relation and rely on the continued (re)enactment of a set of constitutive relations to subsequently act and afford.

Although new materialist and assemblage accounts are interesting, if we take a relational view of networks seriously, this suggests that every entity is ultimately connected to one another. Some writers are happy to embrace such a position. For instance, Vitale (2014: 17) argues, 'it is all networks, all the way down, simply of differing sorts'. From this 'connectivist' (Culp 2016) materialist perspective, any network is composed of objects that are ultimately made of matter and this matter can be broken down into smaller and smaller sub-units (presumably down to the level of quarks). Each object is a network of components, which is a network of substances, that is, a network of atoms and so on. Different from a material understanding of networks, a phase approach suggests that objects are the basic unit of analysis. As we argued in Chapter 2, technical objects are more or less allopoietic and homeostatic things that are not reducible to their relations with other things. From a phase perspective, it is not a matter of drilling down to identify a basic element that is supposedly essential to an object, as the grounding substrate of analysis. Instead, an object is defined as a distinct thing through the qualities it discloses in a given situation. Holding an iPhone in my hand it appears to be a discrete object. But if I pull off the back casing, the casing then appears as a separate discrete object. In the same way if I snap the back casing in half, each piece of the casing appears as a discrete object. Understanding objects as units of appearance rather than assemblages of matter alters the method for understanding how that object perturbs and discloses qualities of other things. Rather than trying to figure out how things are assembled to appear as they do (in the case of the iPhone tracing its network of design, manufacture, shipping and selling), we instead begin by examining the qualities of the iPhone in a given situation and perturbing the iPhone to see how its qualities change as it perturbs and is perturbed by other things.

A material account of networks often considers objects to exist within space, or space to be the outcome of the relations between things (Murdoch 2005). A phase account considers objects as disclosing qualities, through which space and time become intelligible. Here, intelligibility is not just a capacity of any sentient being that perturbs other objects, but is common to all objects. In relation to smart objects, this point is easy to make as many have an intentional and protentional structure that allows them to mimic capacities of sentient beings, such as the ability to detect pressure or temperature. But, we could push this point and state that all objects can contribute to the production of spatial and temporal intelligibility through the way they perturb one another, which in turn discloses particular qualities that shape how space and time appears. Focusing on phases as sets of objects allows us to consider how non-human relations between objects in general are key to shaping human experiences of space and time, even when these relations are not recognized or apprehended by the humans who find themselves living alongside these phases.

In relation to media specifically, an account of phases would therefore question work that introduces a distinction between the digital and physical or virtual and actual to understand the relationship between smart objects and space. As Frith (2015: 7) points out, these distinctions are widespread. In his words, 'While it is possible to look at oppositions of the digital and physical as outdated, strands of that thought survive today. People still use the phrase "in real life" to compare interactions in the physical world to something that happens online … and … Sherry Turkle … still argue[s] that online life is distracting people from the physical world.'

In a similar manner, an account of phases in relation to media would also be opposed to existing work that considers smart objects as producing mixed spaces composed of actual and virtual parts, where smart objects such as locative media are seen as nodes that 'intertwine … the world of atoms with the world of bits' (Frith 2015: 6). For instance, de Souza e Silva (2006, 2006, 2013) sees smartphones as creating hybrid spaces. de Souza e Silva (2006: 265–6) defines hybrid space in the following way: 'A hybrid space … is a conceptual space created by the merging of borders between physical and digital spaces, because of the use of mobile technologies as social devices. Nevertheless, a hybrid space is not constructed by technology. It is built by the connection of mobility and communication and materialized by social networks.' Smart objects do not act as meeting points where different 'forms' of space coincide. This is because from a phase perspective there is no physical, virtual, actual or hybrid space to combine, produce or destroy. There are only objects and perturbations between objects that disclose qualities that create the appearance of spatio-temporal relations. From this position, researchers concerned with the relationship between smart objects and space could focus on space as disclosed through perturbations that are specific to particular objects. This approach is beneficial because it encourages a close analysis of smart objects themselves, without reducing these objects to some field or process that supposedly exists outside of them.

Thinking about media and technical networks more specifically, we can use this point to suggest that media networks are not about the destruction of space and time, as some theorists would suggest. For instance, Heidegger (1982) argues that a key aspect of modern technology is to confuse differences between near and far in order to make everything appear to be close at hand, a process translated into English as de-distancing. A simple example of de-distancing would be the fact that it is quicker and easier to Google a dictionary entry using the network of servers that make up the internet, than it would be to turn around from my desk and pick up a paper dictionary from the shelf behind me. Heidegger sees technologies of de-distancing as problematic because they deskill people until they become dependent on a small number of things that are close to hand, rather than developing the ability to engage with a wide variety of different objects. In a similar vein Meyrowitz (1985) argues media networks create what Moores (2012) calls a placeless culture. The term placeless culture refers to the way media networks break down the distinction between individual places. This argument is based on the idea that as long as you can connect to a network of publicly available information (such as the internet) it does not matter where you are actually located. In turn, places lose their status as a defining factor in shaping everyday interaction and the relationship between distinct places become erased. The concept of phases that I have argued for across the previous chapters has suggested that smart objects produce the appearance, rather than destruction or erasure, of distance. As such, the argument that de-distancing technologies lead to a de-skilling or placelessness is over stated. Rather than thinking in terms of the breakdown of distance, phases ask us to analyse smart objects and objects more generally in terms of how they create the appearance of space and time, instead of assuming that objects change or alter the relationship between pre-existing places located 'in' space and time.

b) Networks are defined by the presence of objects. Phases are defined by the accessibility of objects.

Phases open new ways of understanding how power operates that is distinctive from network accounts of power. Work on networks and objects in general often rely upon ideas of topology to understand how distant things can affect one another. For example, Allen (2016) suggests that networked power is not about the distance between objects, but the presence of objects. In other words, objects can be very far apart from one another, but still have a great effect. In Allen's (2016: 2) words, 'Topology is more about presence than distance; it is intensive rather than extensive, a relational arrangement where power composes the spaces of which it is a part by stretching, folding or distorting relationships to place certain outcomes within or beyond reach.' Take the case of a multinational corporation. While they may have their base of operations in one country, they can still exercise their capacities through an indirect presence in shell and holding companies in other territories in order to avoid paying tax in their main country of business.

Distinct from a topological approach, a phase perspective would suggest that power is not about the presence of objects, but rather about the accessibility of objects. Objects perturb one another, which disclose new or previously disclosed qualities of those objects. In turn it is through the disclosure of these qualities that we understand a change to have occurred. In relation to smart objects, power is not then the ability to overcome distance, but the ability to perturb an object in such a way as to disclose a quality that has a desirable effect for the agent that perturbs that object. Rather than points connected by lines that fold or distort, smart objects point to the fragility and contingency of power. This is because any phase generated by smart objects is always shaped by the perturbations of non-smart objects it exists alongside, such as trees, buildings, cars, roads, hills and so on, which cannot be fully controlled or accounted for by the designers and manufacturers of smart objects.

A phase approach to power and politics should therefore sound a note of optimism regarding smart objects. Writers such as Papacharissi (2010) suggest that smart objects enable new forms of private sphere organized around a networked citizen. This private sphere is presented as a largely negative phenomenon, where 'the citizen entering the networked cocoon of the private sphere is not interested in the civic obligations of the past, and associates citizenship primarily with autonomy … [and] … sustaining a fantasy of total control' (2010: 163). Distinct from Papacharissi and thinkers such as Hu (2015) discussed in Chapter 5, a phase approach suggests that networks are never inherently oppressive in their design and can always be used otherwise. However, this is not the same as saying that networks are value-free or neutral constructions that become oppressive or emancipatory only when used by oppressive or emancipatory groups. Instead, a phase approach argues that objects shape the conditions of possibility for political issues to emerge as issues at all. In this case, politics is about sensitizing users to how objects construct phases and how these phases can be mobilized to produce a public. As we argued in Chapter 6, phase activism can take a number of forms. On one level, activists can alter or modify the objects that make up a phase in order to change the kinds of space and time that appear. On a more extreme level, activists can utilize objects to involute phases in ways that lead to the dismantling of the spaces, times and objects that phases make intelligible.

c) Networks' capacity to change is defined by shifts that occur either internal or external to the network. Phases' ability to change is defined by recurrent causalities between phases and the intelligibilities they disclose.

Beyond a specific reading of smart objects, we can use the concept of phase time to challenge broader kairological accounts of how objects change that have become popular within the social sciences and humanities. For example, work on affect (Massumi 2002; Wetherell, 2012) is often dependent on an explicit temporal theory of relationality to explain how objects can have an effect in the world. Within Deleuzian-inspired theories of affect the encounter is positioned as a kind of event in which the virtuality of a situation is made actual (Anderson 2006). This virtuality and its resulting actuality is not composed by or contained within the objects that make up an encounter, but emerges from the encounter itself, which is anterior to any particular object. This is virtuality as a kind of imperceptible, yet real potential (Dewsbury 2000). Here, 'affect is bound up with the indeterminate movement of spacing and timing … [which] … ties affect to the presence of virtualities that are folded into what has become actual' (Anderson 2006: 737). In doing so, 'movements of affect are always accompanied by a real but virtual knot of tendencies and latencies that generate differences and divergences in what becomes actual' (Anderson 2006: 738). From this perspective, affects are not in any one object and only emerge through the event where objects meet. Here time is presented as Aion, a form of non-metrical excess, which emphasizes the 'vital openness and indeterminacy of experience as it happens' (Dewsbury 2000: 478).

In a similar regard, Harman (2009) argues that ANT approaches also prioritize the encounter as a way of understanding and defining objects as relational deployments that are linked to a particular temporal instance or event. As Harman (2009: 127) puts it, for Latour, 'An object is no more than what it modifies, transforms, perturbs, or creates.' As such, every object is an assemblage or relational outcome of the connection between itself and other things. According to Harman (2009: 128), this results in a situation in which 'Latour does not mind defining an actor by what it affects, but he does not allow an actor to borrow its effects in advance. Payment in real time is demanded at every stage of the translation.' This leads to an account in which the encounter is absolutely fundamental to what an object is and what it can do. Outside of a particular encounter between objects, according to Harman's reading of Latour, time ceases to exist. Here time is theorized as a set of cinematic snapshots or instants, with each encounter creating a temporal frame in which something happens. Such an approach is problematic, because as Harman (2010a: 11) suggests: 'Even the most ardent philosopher of networks would not deny that there is more … than is expressed in a given instant.'

Unlike work on affect or ANT, the concept of phases offers an atemporal account of how objects change. Change does not occur through an external process that happens to an object. Rather, change emerges through recurrent causalities between perturbations that generate phases and the qualities that objects disclose, which in turn modulate how objects can perturb and be perturbed by one another. This account of phases does not return to a simplistic view of a kind of billiard ball causality, where 'units come first, then like billiard balls on a table, they are put into motion and their interactions are the patterns we observe in … life' (Jackson and Nexon 1999: 292). Instead change is fundamentally non-linear, in the sense that it cannot be fully predicted or anticipated in advance, because the qualities disclosed by a perturbation could always be otherwise. The concept of phases then asks us to analyse how space and time come to appear in such and such a way by investigating the perturbations that occur between particular sets of objects that create the appearance of change, rather than seeing change as constituted by a process, system or force that is exterior to the perturbations between those objects.

This non-linear, recurrent account of change and causality offers a very different perspective from work that argues change in relation to smart objects and networks has to be understood contextually. Take the work of writers such as Loader and Mercea (2011), who discuss change and smart objects in relation to politics and power. In their words,

> The more widespread use of social media and internet technologies and their absorption into the mundane practices of lived experience, their potential to shape social relations of power becomes all the greater. Yet,

> such influence is likely to be in ways that are indeterminate and contingent upon a multitude of clashes between social agents, groups and institutions that have competing conceptions of networking democracy. (Loader and Mercea 2011: 759)

For Loader and Marcea, causality is implicitly theorized as dispersed and contingent upon a set of 'social' factors, such as the specificity of social agents and groups that use smart objects. This view is commonplace. We see a similar argument presented in relation to the Arab Spring, when Wolfsfeld et al. (2013: 121) suggest, 'Political change (such as the initial protests associated with the Arab Spring) leads to changes in the use of the social media (e.g., more people signing up and using social media for political content), which can lead to further changes in the political environment (such as more people participating in protests).'

Like Loader and Marcea, for Wolfsfeld et al, smart objects do not themselves cause political change and can only act to catalyse changes that were already set in motion by broader structural or contextual forces. As an alternative to a structural or contextual notion of change and causality, a phase politics of smart objects would understand change as operating through regimes of recurrent causality. Here causality is recurrent in the sense that changes in how people use smart objects can alter the phases and intelligibilities smart objects disclose, which in turn can change the way people engage with these objects in increasingly complex feedback loops. With this point in mind, a phase politics suggests that change and causality can never be reduced to the contextual environment or system in which smart objects are located, but neither is causality reducible to the technical perturbations of the components of smart objects. Instead, accounting for causality becomes a matter of tracing how the qualities of objects appear and identifying which specific human actors and non-human objects instigated these appearances.

Closing remarks

The three differences outlined above are not supposed to offer a manifesto as to why a phase approach for understanding smart objects is better than a point and line model of networks. Instead, by offering these three differences I hope that the concept of phases could help researchers who are interested in understanding technical objects from a non-network perspective as well as providing some starting points for analysing space, time and objects beyond the specific domains of smart and technical things. To reiterate, a phase approach is helpful for two main reasons. First, it allows us to understand how smart objects perturb and are perturbed by other things, which disclose qualities that modulate the appearance of space and time for a variety of humans and non-humans. Second, it provides a methodology for studying smart objects and objects more generally. Focusing on perturbations and qualities enables researchers to stay with objects as the key site where space and time and change and difference are located, without invoking processes or fields outside of these objects. This is useful because concentrating on relations, fields and processes often results in accounts of relations, fields and processes that end up ignoring the very objects they purport to study.

As the diffusion of smart objects become ever more widespread throughout both the global North and South, the density and ubiquity of phases will only become more important in shaping the spatial and temporal intelligibility of humans and non-humans that exist alongside these objects. It is my hope that this book offers a starting point for bringing the mechanisms and logics of phases into the spotlight. It is only once we understand how phases operate and recognize the importance they play in shaping everyday life that an adequate set of political tools can be developed to investigate and respond to these phases and the intelligibilities they modulate.

BIBLIOGRAPHY

Adams, G. (2013), 'Is your TV spying on YOU?', *Daily Mail*. Available online: http://www.dailymail.co.uk/sciencetech/article-2513592/Is-TV-spying-YOU.html (accessed 2 March 2017).

Adorno, T. W., and M. Horkheimer (1997), *Dialectic of Enlightenment*, London: Verso.

Alaimo, S. (2013), 'Thinking as the Stuff of the World', *O-Zone: A Journal of Object-Oriented Studies*, 1: 13–21.

Allen, J. (2016), *Topologies of Power: Beyond Territory and Networks*, London: Routledge Publishing.

Amin, A., and N. Thrift (2013), *Arts of the Political: New Openings for the Left*, Durham: Duke University Press.

Amin, A. and N. Thrift (2017), *Seeing Like a City*, Cambridge: Polity Press.

Amoore, L. (2013), *The Politics of Possibility*, Durham: Duke University Press.

Amoore, L., and V. Piotukh (2015), 'Life beyond big data: Governing with little analytics', *Economy and Society*, 44 (3): 341–66.

Amoore, L., and V. Piotukh, eds (2016), *Algorithmic Life: Calculative Devices in the Age of Big Data*, London: Routledge.

Anderson, B. (2006), 'Becoming and being hopeful: Towards a theory of affect', *Environment and Planning D*, 24 (5): 733.

Anderson, B., and J. Wylie (2009), 'On geography and materiality', *Environment and Planning. A*, 41 (2): 318.

Anderson, B., and C. McFarlane (2011), 'Assemblage and geography', *Area*, 43 (2): 124–7.

Anderson, M., and S. L. Anderson (2011), *Machine Ethics*, Cambridge: Cambridge University Press.

Andrejevic, M. (2007), 'Surveillance in the digital enclosure', *The Communication Review*, 10 (4): 295–317.

Apple. (2015), 'MFI Program', *Apple*. Available online: https://developer.apple.com/programs/mfi/ (accessed 15 September 2015).

Apple. (2017), 'Your heart rate. What it means, and where on Apple Watch you'll find it', *Apple*. Available online: https://support.apple.com/en-gb/HT204666 (accessed 10 March 2017).

Ash, J. (2015a), *The Interface Envelope: Gaming, Technology, Power*, New York: Bloomsbury.

Ash, J. (2015b), 'Sensation, networks, and the GIF: Toward an allotropic account of affect', in K. Hillis, S. Paasonen and M. Petit (eds), *Networked Affect*, 119–35, Cambridge, MA: MIT.

Ash, J., and L. Gallacher (2015), 'Becoming attuned: objects, affects and embodied methodology', in M. Perry and C. Medina (eds), *Methodologies of Embodiment: Inscribing Bodies in Qualitative Research*, 69–85, London: Routledge.

Ash, J., and P. Simpson (2016), 'Geography and post-phenomenology', *Progress in Human Geography*, 40 (1): 48–66.

Ash, J., R. Kitchin and A. Leszczynski (2016), 'Digital turn, digital geographies?', *Progress in Human Geography*, First published date: 24 August 2016, DOI:10.1177/0309132516664800.

Barad, K. (2003), 'Posthumanist performativity: Toward an understanding of how matter comes to matter', *Signs*, 28 (3): 801–31.

Barad, K. (2007), *Meeting the Universe Halfway: Quantum Physics and the Entanglement of Matter and Meaning*, Durham: Duke University Press.

Barbour, I. (2011), 'Philosophy and human values', in M. Winston and R. Edelbach (eds), *Society, Ethics, and Technology*, 113–20, Boston, MA: Cengage Learning.

Bardin, A. (2015), *Epistemology and Political Philosophy in Gilbert Simondon: Individuation, Technics, Social Systems*, Netherlands: Springer.

Bastian, M. (2016), 'Liberating clocks: Developing a critical horology to rethink the potential of clock time', *New Formations*, In Press.

Batty, M. (2013), 'Big data, smart cities and city planning', *Dialogues in Human Geography*, 3 (3): 274–79.

Batty, M., K. W. Axhausen, F., Giannotti, A., Pozdnoukhov, A., Bazzani, M., Wachowicz, G., Ouzounis and Y. Portugali (2012), 'Smart cities of the future', *The European Physical Journal Special Topics*, 214 (1): 481–518.

Beer, D. (2009), 'Power through the algorithm? Participatory web cultures and the technological unconscious', *New Media & Society*, 11 (6): 985–1002.

Beer, D., and R. Burrows (2013), 'Popular culture, digital archives and the new social life of data', *Theory, Culture & Society*, 30 (4): 47–71.

Bell, D. (2016), 'Dyson 360 Eye review: despite what every other reviewer says, really not quite the finished article', *T3*. Available online: http://www.t3.com/reviews/dyson-360-eye-review (accessed 2 April 2016).

Bennett, J. (2009), *Vibrant Matter: A Political Ecology of Things*, Durham: Duke University Press.

Berleur, J. J., and K. Brunnstein (1996), *Ethics of Computing: Codes, Spaces for Discussion and Law*, Cham: Springer Science & Business Media.

Berry, D. M. (2014), *Critical Theory and the Digital*, New York: Bloomsbury Academic.

Bertel, T. F. (2016), '"Why would you want to know?" The reluctant use of location sharing via check-ins on Facebook among Danish youth', *Convergence: The International Journal of Research into New Media Technologies*, 22 (2): 162–76.

Best, K. (2010), 'Living in the control society: Surveillance, users and digital screen technologies', *International Journal of Cultural Studies*, 13 (1): 5–24.

Bimber, B. (1990), 'Karl Marx and the three faces of technological determinism', *Social Studies of Science*, 20 (2): 333–51.

BIS (2013). 'The smart city market: Opportunities for the UK', *Department for Business, Innovation and Skills*, London.

Bissell, D., and V. J. Del Casino (2017), 'Whither labor geography and the rise of the robots?', *Social & Cultural Geography*, 18 (3): 435–42.

Bland, A. (2014), 'FireChat – the messaging app that's powering the Hong Kong protests', *The Guardian*. Available online: http://www.theguardian.com/world/2014/sep/29/firechat-messaging-app-powering-hong-kong-protests (accessed 2 March 2017).

Boer, R. (2013), 'Revolution in the event: The problem of Kairós', *Theory, Culture & Society*, 30 (2): 116–34.

Bogost, I., and N. Montfort (2007), 'New media as material constraint. An introduction to platform studies', *Electronic Techtonics: Thinking at the Interface*, Proceedings of the First International HASTAC Conference: 176–92.

Bonnefon, J.-F., A. Shariff and I. Rahwan (2016), 'The social dilemma of autonomous vehicles', *Science*, 352 (6293): 1573–6.

Bostrom, N., and E. Yudkowsky (2014), 'The ethics of artificial intelligence', in K. Frankish and W. Ramsey (eds), *The Cambridge Handbook of Artificial Intelligence*, 316–34, Cambridge, UK: Cambridge University Press.

Bratich, J. Z. (2006), '"Nothing is left alone for too long" reality programming and control society subjects', *Journal of Communication Inquiry*, 30 (1): 65–83.

Bratton, B. H. (2016), *The Stack: On Software and Sovereignty*, Cambridge, MA: MIT Press.

Brey, P. (2000), 'Disclosive computer ethics', *ACM SIGCAS Computers and Society*, 30 (4): 10–16.

Brey, P. (2005), 'The epistemology and ontology of human-computer interaction', *Minds and Machines*, 15 (3–4): 383–98.

Bryant, L. R. (2011), *The Democracy of Objects*, Ann Arbor: Open Humanities Press.

Bryant, L. R., N. Srnicek and G. Harman (2011), *The Speculative Turn: Continental Materialism and Realism*, Amsterdam: re.Press.

Buchanan, B. (2008), *Onto-Ethologies: The Animal Environments of Uexkull, Heidegger, Merleau-Ponty, and Deleuze*, New York: Suny Press.

Calarco, M. (2008), *Zoographies: The question of the animal from Heidegger to Derrida*, New York: Columbia University Press.

Campbell, R. (1997), 'Philosophy and the accident', in R. Cooter and B. Luckin (eds), *Accidents in History: Injuries, Fatalities and Social Relations*, 17–35, Amsterdam: Rodopi.

Campbell, T. (2013), *Beyond smart cities: how cities network, learn and innovate*, London: Routledge.

Cannon, W. (1929), 'Organization for physiological homeostasis', *Physiological Reviews*, 9: 399–431.

Chao, R. (2014), 'FireChat wasn't meant for protests. Here's how it worked (Or Didn't) at occupy central', *Tech President*. Available online: http://techpresident.com/news/25304/firechat-wasn%E2%80%99t-meant-protests-here%E2%80%99s-how-it-worked-or-didn%E2%80%99t-occupy-central (accessed 2 March 2017).

Che, J. (2015), '9 countries that aren't giving uber an inch', *Huffington Post*. Available online: http://www.huffingtonpost.com/entry/uber-countries-governments-taxi-drivers_us_55bfa3a9e4b0d4f33a037a4b (accessed 2 March 2017).

Chipworks. (2012), 'Systems analysis of the apple lightning to USB cable', *Chipworks*. Available online: http://www.chipworks.com/about-chipworks/overview/blog/systems-analysis-apple-lightning-usb-cable (accessed 2 September 2016).

Chun, W. H. K. (2011), *Programmed Visions: Software and Memory*, Cambridge, MA: MIT Press.

Clarke, P. (2015), 'When the internet of things and smart machines collide', *Wired*. Available online: http://www.wired.com/insights/2015/03/internet-things-smart-machines-collide/(accessed 2 March 2017).

Coleman, R. (2015), *Transforming images: screens, affect, futures*, London: Routledge.

Coley, R., and D. Lockwood (2012), *Cloud Time: The Inception of the Future*, Winchester: Zero Books.

Connolly, W. E. (2013), 'The "New Materialism" and the Fragility of Things', *Millennium – Journal of International Studies*, 41 (3): 399–412.

Coole, D., and S. Frost (2010), 'Introducing the new materialisms', in D. Coole and S. Frost (eds), *New Materialisms: Ontology, Agency, and Politic*, 1–47, Durham: Duke University Press.

Couldry, N. (2012), *Why Voice Matters: Culture and Politics after Neoliberalism*, London: Sage Publishing.

Crampton, J. W., M. Graham, A., Poorthuis, T., Shelton, M., Stephens, M., W. Wilson and M. Zook (2013), 'Beyond the geotag: Situating "big data" and leveraging the potential of the geoweb', *Cartography and Geographic Information Science*, 40 (2): 130–9.

Crang, M., and S. Graham (2007), 'Sentient cities: Ambient intelligence and the politics of urban space', *Information, Communication and Society*, 10 (6): 789–817.

Crisp, R. (1995), 'Deontological ethics', in T. Honderich (ed.), *The Oxford Companion to Philosophy*, 187–8, New York: Oxford University Press.

Crockett, C. (2012), 'Foreword', in C. Malabou (ed.), *Plasticity at the Dusk of Writing: Dialectic, Destruction, Deconstruction*, Columbia: Columbia University Press.

Culp, A. (2016), *Dark Deleuze*, Minneapolis: University of Minnesota Press.

Davies, W. (2014), *The Limits of Neoliberalism: Authority, Sovereignty and the logic of competition*, New York: Sage Publishing.

Davis, A. (2014), 'Hong Kong Court Bans Occupations as Police Warn of Riot', *Bloomberg*. Available online: http://www.bloomberg.com/news/articles/2014-10-20/hong-kong-court-bans-street-occupation-as-police-warn-of-riot- (accessed 2 March 2017).

DCLC (2001), 'Planning policy guidance note 8: Telecommunications', *Department for Communities and Local Government*. Available online: http://www.medway.gov.uk/pdf/telecommunications_ppg8.pdf (accessed 2 March 2017).

De Boever, A., A. Murray, J. Roffe and A. Woodward (2012), *Gilbert Simondon: Being and Technology*, Edinburgh: Edinburgh University Press.

de Souza e Silva, A. (2006), 'From cyber to hybrid mobile technologies as interfaces of hybrid spaces', *Space and Culture*, 9 (3): 261–78.

de Souza e Silva, A. (2013), 'Location-aware mobile technologies: Historical, social and spatial approaches', *Mobile Media & Communication*, 1 (1): 116–21.

de Souza e Silva, A., and G. C. Delacruz (2006), 'Hybrid reality games reframed: potential uses in educational contexts', *Games and Culture*, 1 (3): 231–51.

Dean, J. (2005), 'Communicative capitalism: Circulation and the foreclosure of politics', *Cultural Politics*, 1 (1): 51–74.

Del Lucchese, F. (2009), 'Monstrous individuations: Deleuze, simondon, and relational ontology', *Differences*, 20 (2–3): 179–93.

DeLanda, M. (2016), *Assemblage Theory*, Edinburgh: Edinburgh University Press.

Deleuze, G. (1988a), *Bergsonism*, New York: Zone Books.

Deleuze, G. (1988b), *Spinoza: Practical Philosophy*, San Francisco: City Lights Publishers.

Deleuze, G. (1992), 'Postscript on the societies of control', *October* 59 (Winter): 3–7.

Deleuze, G., and F. Guattari (1987), *A Thousand Plateaus: Capitalism and Schizophrenia*, Minneapolis: University of Minnesota Press.

Derrida, J. (1998), *Of Grammatology*, Baltimore: Johns Hopkins University Press.

Dewsbury, J.-D. (2000), 'Performativity and the event: Enacting a philosophy of difference', *Environment and Planning D: Society and Space*, 18 (4): 473–96.

Dodge, M., and R. Kitchin (2005), 'Codes of life: Identification codes and the machine readable world', *Environment and Planning D: Society & Space*, 23 (6): 851–81.

Dodge, M., and R. Kitchin (2007), '"Outlines of a world coming into existence": pervasive computing and an ethics of forgetting', *Environment and Planning B: Planning and Design*, 34 (3): 431–45.

Dodge, M., and R. Kitchin (2009), 'Software, objects, and home space', *Environment and Planning A*, 41 (6): 1344–65.

Dodge, M., R. Kitchin and M. Zook (2009), 'How does software make space? Exploring some geographical dimensions of pervasive computing and software studies', *Environment and Planning A*, 41 (6): 1283–93.

Drake, F. (2006), 'Mobile phone masts: Protesting the scientific evidence', *Public Understanding of Science*, 15 (4): 387–410.

Drake, F. (2010), 'Protesting mobile phone masts: Risk, neoliberalism, and governmentality', *Science, Technology & Human Values*, 36 (4): 522–48.

Dreyfus, H. L. (1991), *Being-in-the-world: A Commentary on Heidegger's Being and Time, Division I*, Cambridge, MA: MIT Press.

Durham Peters, J. (2015), *The Marvelous Clouds: Towards a Philosophy of Elemental Media*, Chicago: University of Chicago Press.

Dyson. (2017), '360 Eye Engineering Story', *Dyson*. Available online: http://www.dyson.com/vacuum-cleaners/robot/dyson-360-eye/engineering-story.aspx (accessed 2 March 2017).

Editorials. (2014), 'Protesters who blocked roads also cleared Hong Kong's polluted air' *South China Morning Post*. Available online: http://www.scmp.com/comment/insight-opinion/article/1612281/protesters-who-blocked-roads-also-cleared-hong-kongs (accessed 2 March 2017).

Edmonds, D. (2013), *Would you Kill the Fat Man?: The Trolley Problem and what your Answer Tells us about Right and Wrong*, Princeton, NJ: Princeton University Press.

Elwood, S. (2006), 'Beyond cooptation or resistance: Urban spatial politics, community organizations, and GIS-Based spatial narratives', *Annals of the Association of American Geographers*, 96 (2): 323–41.

Elwood, S., and A. Leszczynski (2011), 'Privacy, reconsidered: New representations, data practices, and the geoweb', *Geoforum*, 42 (1): 6–15.

Engelmann, S. (2015), 'Toward a poetics of air: Sequencing and surfacing breath', *Transactions of the Institute of British Geographers*, 40 (3): 430–44.

Ernst, W. (2013), *Digital Memory and the Archive*, Minneapolis: University of Minnesota Press.

Farman, J. (2012), *Mobile Interface Theory: Embodied Space and Locative Media*, London: Routledge Publishing.

Felski, R. (2015), *The Limits of Critique*, Chicago: Chicago University Press.

Fleming, C. (2016), 'Tesla car mangled in fatal crash was on Autopilot and speeding, NTSB says', *LA Times*. Available online: http://www.latimes.com/business/autos/la-fi-hy-autopilot-photo-20160726-snap-story.html (accessed 2 March 2017).

Foggo, D. (2003), 'Protesters topple mobile phone masts as health scare spreads', *The Telegraph*. Available online: http://www.telegraph.co.uk/news/uknews/1448109/Protesters-topple-mobile-phone-masts-as-health-scare-spreads.html (accessed 2 April 2016).

Foot, P. (1967), 'The problem of abortion and the doctrine of double effect', *Oxford Review* 5: 1–6.

Franklin, S. (2015), *Control: Digitality as Cultural Logic*, Cambridge, MA: MIT Press.

Fraser, M., S. Kember and C. Lury (2005), *Inventive life: Approaches to the new vitalism*, Thousand Oaks, CA: Sage Publications.

Frith, J. (2013), 'Turning life into a game: Foursquare, gamification, and personal mobility', *Mobile Media & Communication*, 1 (2): 248–62.

Frith, J. (2015), *Smartphones as locative media*, London: Polity Press.

Fuchs, C. (2012), 'Social media, riots, and revolutions', *Capital & Class*, 36 (3): 383–91.

Fuller, M. (2003), *Behind the blip: Essays on the culture of software*, New York, NY: Autonomedia.

Fuller, M. (2008), *Software Studies: A Lexicon*, Cambridge, MA: MIT Press.

Gabel, V., M. Maire, C. F. Reichert, S. L. Chellappa, C. Schmidt, V. Hommes, A. U. Viola and C. Cajochen (2013), 'Effects of artificial dawn and morning blue light on daytime cognitive performance, well-being, cortisol and melatonin levels', *Chronobiology International*, 30 (8): 988–97.

Gabrys, J. (2014), 'Programming environments: Environmentality and citizen sensing in the smart city', *Environment and Planning D: Society and Space*, 32 (1): 30–48.

Gabrys, J. (2015), *Program Earth: Environmental Sensing Technology and the Making of a Computational Planet*, Minneapolis: University of Minnesota Press.

Gabrys, J. (2016), 'Practicing, materialising and contesting environmental data', *Big Data & Society*, 3 (2): 1–7.

Gallagher, M. (2016), 'Sound as affect: Difference, power and spatiality', *Emotion, Space and Society*, 20: 42–8.

Gallagher, S., and D. Zahavi (2012), *The Phenomenological Mind*, London: Routledge.

Galloway, A. (2004), 'Intimations of everyday life: Ubiquitous computing and the city', *Cultural studies*, 18 (2–3): 384–408.

Galloway, A. R. (2004), *Protocol: How control exists after decentralization*, Cambridge, MA: MIT Press.

Galloway, A. R., and E. Thacker (2007), *The Exploit: A Theory of Networks*, Minneapolis, MN: University of Minnesota Press.

Garde-Hansen, J. (2011), *Media and Memory*, Edinburgh: Edinburgh University Press.

Gazzard, A. (2011), 'Location, location, location: Collecting space and place in mobile media', *Convergence: The International Journal of Research into New Media Technologies*, 17 (4): 405–17.

Gene. (2017), 'Inside Tesla's autopilot 2.0 roll out: release notes and screenshots', *Teslerati*. Available online: http://www.teslarati.com/tesla-extends-lock-period-free-unlimited-supercharging/ (accessed 2 March 2017).

Gerlach, J., and T. Jellis (2015), 'Guattari: Impractical philosophy', *Dialogues in Human Geography*, 5 (2): 131–48.

Gilbert, M. (2010), 'Theorizing digital and urban inequalities: Critical geographies of "race," gender and technological capital', *Information, Communication & Society*, 13 (7): 1000–18.

Gilmore, J. N. (2016), 'Everywear: The quantified self and wearable fitness technologies', *New Media & Society*, 18 (11): 2524–39.

Glendinning, S. (1996), 'Heidegger and the question of animality', *International Journal of Philosophical Studies*, 4 (1): 67–86.

Goggin, G. (2012). *Cell phone culture: Mobile technology in everyday life*, London: Routledge.

Goggin, G., and L. Hjorth (2014), *The Routledge Companion to Mobile Media*, London: Routledge.

Golson, J. (2017), 'Read the Florida highway patrol's full investigation into the fatal Tesla crash', *The Verge*. Available online: http://www.theverge.com/2017/2/1/14458662/tesla-autopilot-crash-accident-florida-fatal-highway-patrol-report (accessed 2 March 2017).

Gontarski, S. (2012), 'Creative involution: Bergson, beckett, deleuze', *Deleuze Studies*, 6 (4): 601–13.

Goodall, N. (2016), 'Can You Program Ethics Into a Self-Driving Car?' *Spectrum IEEE*. Available online: http://spectrum.ieee.org/transportation/self-driving/can-you-program-ethics-into-a-selfdriving-car (accessed 1 March 2017).

Goodall, N. J. (2014), 'Machine ethics and automated vehicles', in G. Meyer and S. Beiker (eds), *Road Vehicle Automation*, 93–102, London: Springer.

Google (2016), 'Google self-driving car project monthly report', *Google*. Available online: https://static.googleusercontent.com/media/www.google.com/en//selfdrivingcar/files/reports/report-0216.pdf (accessed 7 January 2017).

Department for Communities, and Local Government (2012), 'National Planning Policy Framework'. *British Government*. Available online: https://www.gov.uk/government/publications/national-planning-policy-framework-2 (accessed 1 March 2017).

Grabianowski, E. (2006), 'How speech recognition works', *How Stuff Works*. Available online: http://electronics.howstuffworks.com/gadgets/high-tech-gadgets/speech-recognition1.htm (accessed 12 December 2015).

Graham, M. (2017), 'Digitally augmented geographies', in R. Kitchin, T. P. Lauriault and M. W. Wilson (eds), *Understanding Spatial Media*, 44–55, London: Sage.

Graham, S., and S. Marvin (2001), *Splintering Urbanism: Networked Infrastructures, Technological Mobilities and the Urban Condition*, London: Psychology Press.

Greenfield, A. (2006), *Everyware: The dawning age of ubiquitous computing*, Berkeley, CA: New Riders.

Greengard, S. (2015), *The Internet of Things*, Cambridge, MA: MIT Press.

Gregg, M., and G. J. Seigworth (2010), *The Affect Theory Reader*, Durham: Duke University Press.

Hansen, M. (2000a), 'Becoming as creative involution?: Contextualizing Deleuze and Guattari's biophilosophy', *Postmodern Culture*, 11 (1): n.p.

Hansen, M. (2000b), *Embodying Technesis: Technology Beyond Writing*, Chicago: University of Michigan Press.
Hansen, M. (2004), *New Philosophy for New Media*, Cambridge, MA: MIT Press.
Hansen, M. (2006a), *Bodies in code: Interfaces with digital media*, New York: Routledge.
Hansen, M. (2006b), 'Media theory', *Theory, Culture & Society*, 23 (2–3): 297–306.
Hansen, M. (2009), 'Living (with) technical time: From media surrogacy to distributed cognition', *Theory, Culture & Society*, 26 (2–3): 294–315.
Hansen, M. (2015), *Feed Forward: On the future of twenty first century media*, Chicago: Chicago University Press.
Harman, G. (2005), *Guerrilla Metaphysics: Phenomenology and the Carpentry of Things*, Illinois: Open Court.
Harman, G. (2009), *Prince of Networks: Bruno Latour and Metaphysics*, Amsterdam: re. Press.
Harman, G. (2010a), *Circus Philosophicus*, Hants: O Books.
Harman, G. (2010b), *Towards Speculative Realism: Essays and Lectures*, Winchester: Zero Books.
Harman, G. (2013), *Bells and Whistles: More Speculative Realism*, Winchester: Zero Books.
Harman, G. (2014), *Bruno Latour: Reassembling the Political*, London: Pluto Press.
Harman, G. (2016), *Immaterialism*, Cambridge: Polity Press.
Harvey, D. (1992), *The Condition of Postmodernity: An Enquiry into the Origins of Cultural Change*, London: Wiley.
Hayles, N. K. (1999), *How we became posthuman: virutal bodies, in cybernetics, literature and informatics*, Cambridge, MA: MIT Press.
Hayles, N. K. (2009), 'RFID: Human agency and meaning in information-intensive environments', *Theory, Culture & Society*, 26 (2–3): 47–72.
Hayles, N. K. (2012), *How We Think: Digital Media and Contemporary Technogenesis*, Chicago: University of Chicago Press.
Heidegger, M. (1962), *Being and Time*, Oxford, UK: Blackwell Publishing.
Heidegger, M. (1982), *The Question Concerning Technology, and Other Essays*, London: HarperCollins.
Heidegger, M. (1992a), *The Concept of Time* Oxford, UK: Blackwell Publishing.
Heidegger, M. (1992b), *History of the Concept of Time*, Bloomington, IN: Indiana University Press.
Heidegger, M. (1995), *The Fundamental Concepts of Metaphysics: World, Finitude, Solitude*, Bloomington, IN: Indiana University Press.
Heil, J. (2004), 'Powers: A study in metaphysics. George Molnar', *The Journal of Philosophy*, 101 (8): 438–43.
Hevelke, A., and J. Nida-Rümelin (2015), 'Responsibility for crashes of autonomous vehicles: An ethical analysis', *Science and Engineering Ethics*, 21 (3): 619–30.
Howells, C. (2013), *Derrida: Deconstruction from phenomenology to ethics*, London: John Wiley & Sons.
Hu, T. (2015), *A Prehistory of the Cloud*, Cambridge, MA: MIT Press.
Hui, Y. (2016), *On the Existence of Digital Objects*, Minnesota: University of Minnesota Press.

Humphreys, L. (2010), 'Mobile social networks and urban public space', *New Media & Society*, 12 (5): 763–78.

Hunn, N. (2015), 'The Market for Smart Wearable Technology: A Consumer Centric Approach', *WiFore Wireless Consulting*. Available online: http://www.nickhunn.com/wp-content/uploads/downloads/2014/08/The-Market-for-Smart-Wearables.pdf (accessed 1 March 2017).

Husserl, E. (1991), *On the phenomenology of the consciousness of internal time (1893-1917)*, Dordrecht: Kluwer Academic.

Hutchins, E. (1996), *Cogntion in the Wild*, Cambridge, MA: MIT Press.

Hwang, T., and M. Elish. (2015), 'The mirage of the marketplace', *Slate*. Available online: http://www.slate.com/articles/technology/future_tense/2015/07/uber_s_algorithm_and_the_mirage_of_the_marketplace.html (accessed 1 March 2017).

Ihde, D. (2004), 'Philosophy of technology', in P. Kemper (ed.), *Philosophical Problems Today*, 91–108, Berlin: Springer.

Ihde, D. (2010), *Heidegger's Technologies: Postphenomenological Perspectives*, New York: Fordham University Press.

Iliadis, A. (2015), 'Mechanology: Machine typologies and the birth of philosophy of technology in France (1932-1958)', *Systema*, 3(1): 131–44.

Introna, L. D. (2005), 'Disclosive ethics and information technology: Disclosing facial recognition systems', *Ethics and Information Technology*, 7 (2): 75–86.

Introna, L. D. (2007), 'Maintaining the reversibility of foldings: Making the ethics (politics) of information technology visible', *Ethics and Information Technology*, 9 (1): 11–25.

Jackman, A. H. (2016), 'Rhetorics of possibility and inevitability in commercial drone tradescapes', *Geographica Helvetica*, 71 (1): 1–6.

Johansson, A., S. Nordin, M. Heiden and M. Sandström (2010), 'Symptoms, personality traits, and stress in people with mobile phone-related symptoms and electromagnetic hypersensitivity', *Journal of Psychosomatic Research*, 68 (1): 37–45.

Jones, M. (2009), 'Phase space: Geography, relational thinking, and beyond', *Progress in Human Geography*, 33 (4): 487–506.

Judah, S. (2014), 'Hong Kong's "off-grid" protesters', *BBC*. Available online: http://www.bbc.co.uk/news/blogs-trending-29411159 (accessed 1 March 2017).

Kanngieser, A. (2012), 'A sonic geography of voice: Towards an affective politics', *Progress in Human Geography*, 36 (3): 336–53.

Karppi, T. (2015), 'Happy accidents: Facebook and the value of affect', in K. Hillis, S. Paasonen and M. Petit (eds), *Networked Affect*, 221–35, Cambridge, MA: MIT.

Kember, S., and J. Zylinska (2015), *Life After New Media: Mediation as Vital Process*, Cambridge, MA: MIT Press.

Khondker, H. H. (2011), 'Role of the new media in the Arab Spring', *Globalizations*, 8 (5): 675–79.

Khosla, E. (2015), 'Here's everywhere Uber is banned around the world', *Business Insider*. Available online: http://www.businessinsider.com/heres-everywhere-uber-is-banned-around-the-world-2015-4?IR=T (accessed 1 March 2017).

Kinsley, S. (2011), 'Anticipating ubiquitous computing: Logics to forecast technological futures', *Geoforum*, 42 (2): 231–40.

Kinsley, S. (2012), 'Futures in the making: Practices to anticipate "ubiquitous computing"', *Environment and Planning A*, 44 (7): 1554–69.

Kinsley, S. (2015), 'Memory programmes: The industrial retention of collective life', *Cultural Geographies*, 22 (1): 155–75.

Kirschenbaum, M. G. (2008), *Mechanisms: New Media and the Forensic Imagination*, Cambridge, MA: MIT Press.

Kitchin, R. (2011), 'The programmable city', *Environment and Planning B: Planning and Design*, 38 (6): 945–51.

Kitchin, R. (2014a), *The data revolution: Big data, open data, data infrastructures and their consequences*, London: Sage Publishing.

Kitchin, R. (2014b), 'The real-time city? Big data and smart urbanism', *GeoJournal*, 79 (1): 1–14.

Kitchin, R., and M. Dodge (2011), *Code/Space: Software and Everyday Life*, Cambridge, MA: MIT Press.

Kroes, P., and A. Meijers (2006), 'The dual nature of technical artefacts', *Studies in History and Philosophy of Science Part A*, 37 (1): 1–4.

Kuniavsky, M. (2010), *Smart Things: Ubiquitous Computing User Experience Design*, Burlington, MA: Elsevier.

Kyrre, J., Olsen, B., Selinger, E. and Riis, S. eds (2008), *New Waves in Philosophy of Technology*, London: Palgrave Macmillan.

Lachlan, R. (1893), *An elementary treatise on modern pure geometry*, London: Macmillan and Company.

LaMarre, T. (2013), 'Afterword: Humans and machines', in M. Combes (ed.), *Gilbert Simondon and the Philosophy of the Transindividual*, 79–109, Cambridge, MA: MIT Press.

Lapworth, A. (2015), 'Habit, art, and the plasticity of the subject: The ontogenetic shock of the bioart encounter', *Cultural Geographies*, 22 (1): 85–102.

Lapworth, A. (2016), 'Theorizing bioart encounters after Gilbert Simondon', *Theory, Culture & Society*, 33: 123–50.

Latour, B. (1996), *Aramis, or, The Love of Technology*, Cambridge, MA: Harvard University Press.

Latour, B. (2004a), *Politics of Nature*, Cambridge, MA: Harvard University Press.

Latour, B. (2004b), 'Why has critique run out of steam? From matters of fact to matters of concern', *Critical Inquiry*, 30 (2): 225–48.

Latour, B. (2005), *Reassembling the Social: An Introduction to Actor-Network-Theory*, Oxford: Oxford University Press.

Latour, B. (2007), 'Turning around politics: A note on Gerard de Vries' paper', *Social Studies of Science*, 37 (5): 811–20.

Latour, B. (2013), *An Inquiry into Modes of Existence*, Cambridge, MA: Harvard University Press.

Law, J. (2002), 'Objects and spaces', *Theory, Culture & Society*, 19 (5–6): 91–105.

Law, J. (2004), *After Method: Mess in Social Science Research*, London: Taylor & Francis.

Lea, J. J. (2009), 'Post-phenomenology/post-phenomenological geographies', in R. Kitchin and N. Thrift (eds), *International Encyclopedia of Human Geography*, 373–8, Oxford: Elsevier.

Lee, A. Y. L., and K. W. Ting (2015), 'Media and information praxis of young activists in the Umbrella movement', *Chinese Journal of Communication*, 8 (4): 376–92.

Lee, F. L., and J. M. Chan (2016), 'Digital media activities and mode of participation in a protest campaign: A study of the Umbrella Movement', *Information, Communication & Society*, 19 (1): 4–22.

Leszczynski, A. (2014), 'Spatial media/tion', *Progress in Human Geography*, 39 (6): 729–51.

Licoppe, C. (2013), 'Merging mobile communication studies and urban research: Mobile locative media, "onscreen encounters" and the reshaping of the interaction order in public places', *Mobile Media & Communication*, 1 (1): 122–8.

Lin, P. (2013), 'The Ethics of Autonomous Cars', *The Atlantic*. Available online: https://http://www.theatlantic.com/technology/archive/2013/10/the-ethics-of-autonomous-cars/280360/ (accessed 1 March 2017).

Lin, P. (2016), 'Why ethics matters for autonomous cars', in M. Maurer, J. Christian Gerdes, B. Lenz and H. Winner (eds), *Autonomous Driving: Technical, Legal and Social Aspects*, 69–85, Berlin: Springer.

Lin, P., K. Abney and G. A. Bekey (2011), *Robot Ethics: The Ethical and Social Implications of Robotics*, Cambridge, MA: MIT press.

Lirmson, C. (2016), 'Report of Traffic Accident Involving an Autonomous Vehicle. California, Department of Motor Vehicles', *Department of Motor Vehicles*. Available online: https://www.dmv.ca.gov/portal/wcm/connect/3946fbb8-e04e-4d52-8f80-b33948df34b2/Google+Auto+LLC+02.14.16.pdf?MOD=AJPERES (accessed 1 March 2017).

Loader, B. D., and D. Mercea (2011), 'Networking Democracy?', *Information, Communication & Society*, 14 (6): 757–69.

Luhmann, N. (1986), 'The autopoiesis of social systems', *Sociocybernetic Paradoxes*, 172–92.

Luhmann, N. (1995), *Social systems*, Stanford: Stanford University Press.

LumoRun. (2017), 'Science of Lumo run', *Lumo Run*. Available online http://www.lumobodytech.com/science-of-lumo-run/ (accessed 2 March 2017).

Lupton, D. (2014), 'Apps as artefacts: Towards a critical perspective on mobile health and medical apps', *Societies*, 4 (4): 606.

Lupton, D. (2015), 'Quantified sex: A critical analysis of sexual and reproductive self-tracking using apps', *Culture, Health & Sexuality*, 17 (4): 440–53.

Lupton, D. (2016), 'The diverse domains of quantified selves: Self-tracking modes and dataveillance', *Economy and Society*, 45 (1): 101–22.

Mackenzie, A. (2010), *Wirelessness*, Cambridge, MA: MIT Press.

Malabou, C. (2002), *Plasticity at the Dusk of Writing: Dialectic, Destruction, Deconstruction*, New York: Columbia University Press.

Malabou, C. (2004), *The Future of Hegel: Plasticity, Temporality and Dialectic*, Abingdon: Taylor & Francis.

Manders-Huits, N. (2011), 'What values in design? The challenge of incorporating moral values into design', *Science and Engineering Ethics*, 17 (2): 271–87.

Manovich, L. (2013), *Software Takes Command*, New York: Bloomsbury.

Markoff, J. (2016), 'Should your driverless car hit a pedestrian to save your life?', *New York Times*. Available online: https://http://www.nytimes.com/2016/06/24/technology/should-your-driverless-car-hit-a-pedestrian-to-save-your-life.html (accessed 1 March 2017).

Marres, N. (2007), 'The issues deserve more credit: Pragmatist contributions to the study of public involvement in controversy', *Social Studies of Science*, 37 (5): 759–80.

Marshall, K. P. (1999), 'Has technology introduced new ethical problems?', *Journal of Business Ethics*, 19 (1): 81–90.

Masana, M. I., and M. L. Dubocovich (2001), 'Melatonin receptor signaling: finding the path through the dark', *Sci. STKE*, 107: pe39.

Massumi, B. (2002), *Parables for the Virtual: Movement, Affect, Sensation*, Durham: Duke University Press.

Maturana, H. R., and F. J. Varela (1991), *Autopoiesis and Cognition: The Realization of the Living*, Berlin: Springer Science & Business Media.

McCormack, D. P. (2017), 'Elemental infrastructures for atmospheric media: On stratospheric variations, value and the commons', *Environment and Planning D: Society and Space*, 35 (3): 418–37.

McCullough, M. (2004), *Digital Ground: Architecture, Pervasive Computing and Environmental Knowing*, Cambridge, MA: MIT Press.

McCullough, M. (2013), *Ambient Commons: Attention in the Age of Embodied Information*, Cambridge, MA: MIT Press.

Meyrowitz, J. (1985), *No sense of place: The impact of electronic media on social behavior*, Oxford: Oxford University Press.

Michels, J., A. Saxena and A. Y. Ng (2005), 'High speed obstacle avoidance using monocular vision and reinforcement learning', *Proceedings of the 22nd International Conference on Machine Learning*, ACM: n.p.

Mobile Mast (2013), 'Base Stations and Masts.' *Mobile Mast Info*. Available online: http://www.mobilemastinfo.com/base-stations-and-masts/ (accessed 10 March 2016).

Moore, P., and A. Robinson (2016), 'The quantified self: What counts in the neoliberal workplace', *New Media & Society*, 18 (11): 2774–92.

Moores, S. (2012), *Media, Place and Mobility*, London: Palgrave Macmillan.

Morton, T. (2010), *The Ecological Thought*, Cambridge, MA: Harvard University Press.

Mu, C. (2014), 'Firechat – the app connecting Hong Kong protesters', *DW*. Available online: http://www.dw.com/en/firechat-the-app-connecting-hong-kong-protesters/a-17967380 (accessed 1 March 2017).

Müller, M., and C. Schurr (2016), 'Assemblage thinking and actor-network theory: Conjunctions, disjunctions, cross-fertilisations', *Transactions of the Institute of British Geographers*, 41 (3): 217–29.

Murdoch, J. (2005), *Post-structuralist Geography: A Guide to Relational Space*, London: Sage Publications.

Nancy, J. L., and A. Barrau (2015), *What's These Worlds Coming To?*, New York: Fordham University Press.

Nest. (2017a), 'How to set up and use Activity Zones', *Nest*. Available online: https://nest.com/uk/support/article/What-are-Activity-Zones-and-how-do-I-create-them (accessed 1 March 2017).

Nest. (2017b), 'Nest cam indoor', *Nest*. Available online: https://nest.com/uk/camera/meet-nest-cam/ (accessed 1 March 2017).

Neves, M. (2001), 'From the autopoiesis to the allopoiesis of law', *Journal of Law and Society*, 28 (2): 242–64.

Nyholm, S., and J. Smids (2016), 'The ethics of accident-algorithms for self-driving cars: An applied trolley problem?', *Ethical Theory and Moral Practice*, 19 (5): 1275–89.

O'Connor, E. (2007), 'My story', *Radiation Research*. Available online: http://www.radiationresearch.org/pdfs/eileen_my_story.pdf. (accessed 10 December 2015).

Oremus, W. (2016), 'The Tesla Autopilot Crash Victim Was Apparently Watching a Movie When He Died', *Slate*. Available online: http://www.slate.com/blogs/moneybox/2016/07/01/tesla_autopilot_crash_victim_joshua_brown_was_watching_a_movie_when_he_died.html (accessed 1 March 2017).

Out, B. I. (2004), 'Phone Masts – A Health Risk?', *BBC*. Available online: http://www.bbc.co.uk/insideout/westmidlands/series6/phone_masts.shtml (accessed 13 December 2015).

Özkul, D., and L. Humphreys (2015), 'Record and remember: Memory and meaning-making practices through mobile media', *Mobile Media & Communication*, 3 (3): 351–65.

Palladino, V. (2016), 'Lumo run reviewed: Messing up your form? This running coach can fix that', *ArsTechnica*. Available online: https://arstechnica.co.uk/gadgets/2016/08/lumo-run-sensor-review/ (accessed 1 March 2017).

Pantzar, M., and M. Ruckenstein (2015), 'The heart of everyday analytics: Emotional, material and practical extensions in self-tracking market', *Consumption Markets & Culture*, 18 (1): 92–109.

Papacharissi, Z. (2010), *A Private Sphere: Democracy in a Digital Age*, London: Polity Press.

Parrika, J. (2008), 'Copy', in M. Fuller (ed.), *Software Studies: A Lexicon*, 70–8, Cambridge, MA: MIT Press.

Parikka, J. (2012), 'New materialism as media theory: Medianatures and dirty matter', *Communication and Critical/Cultural Studies*, 9 (1): 95–100.

Parikka, J. (2015), *A Geology of Media*, Minneapolis: Minnesota University Press.

Pedwell, C. (2014), *Affective relations: the transnational politics of empathy*, London: Palgrave.

Pink, S., and V. Fors (2017), 'Being in a mediated world: Self-tracking and the mind–body–environment', *Cultural Geographies*, First published date: 12 January 2017, DOI:10.1177/1474474016684127.

Pink, S., and K. Leder Mackley (2013), 'Saturated and situated: Expanding the meaning of media in the routines of everyday life', *Media, Culture & Society*, 35 (6): 677–91.

Pöppel, E., and T. T. Artin (1988), *Mindworks: Time and Conscious Experience*, London: Harcourt.

Raunig, G. (2016), *Dividuum: Machinic Capitalism and Molecular Revolution*, Los Angeles, CA: Semiotext(e).

Reese, H. (2016), 'Autonomous driving levels 0 to 5: Understanding the differences', *Tech Republic*. Available online: http://www.techrepublic.com/article/autonomous-driving-levels-0-to-5-understanding-the-differences/ (accessed 1 March 2017).

Rhodes, M. (2016), 'Sun: A seriously beautiful weather app powered by dark sky', *Wired*. Available online: https://http://www.wired.com/2016/11/sun-seriously-beautiful-weather-app-powered-dark-sky/ (accessed 1 March 2017).

Richardson, L. (2016), 'Feminist geographies of digital work', *Progress in Human Geography*, First published date: 14 November 2016, DOI:10.1177/0309132516677177.

Roberts, T. (2012), 'From "new materialism" to "machinic assemblage": Agency and affect in IKEA', *Environment and Planning A*, 44 (10): 2512–29.

Roberts, W. A. (2002), 'Are animals stuck in time?', *Psychological Bulletin*, 128 (3): 473.

Rogers, B. (2015), 'The social costs of uber', *University of Chicago Law Review*, 82: 85–102.

Rose, G. (2015), 'Rethinking the geographies of cultural "objects" through digital technologies interface, network and friction', *Progress in Human Geography*, 40 (3): 334–51.

Rosenblat, A., and L. Stark (2015), 'Uber's drivers: Information asymmetries and control in dynamic work', *International Journal of Communication*, 10 (27): 1–27.

Rubin, G. J., A. J. Cleare and S. Wessely (2008), 'Psychological factors associated with self-reported sensitivity to mobile phones', *Journal of Psychosomatic Research*, 64 (1): 1–9.

Rubin, G. J., J. D. Munshi and S. Wessely (2005), 'Electromagnetic hypersensitivity: a systematic review of provocation studies', *Psychosomatic Medicine*, 67 (2): 224–32.

Saffron. (2017a), 'Drift FAQ', *Drift*. Available online: https://drift-light.com/#faq (accessed 1 March 2017).

Saffron. (2017b), 'What happened to sleep', *Drift*. Available online: https://drift-light.com (accessed 1 March 2017).

Saker, M., and L. Evans (2016), 'Locative mobile media and time: Foursquare and technological memory', *First Monday*, 21 (2): n.p.

Savage, P. (1995), 'Designing a GUI for business telephone users', *Interactions*, 2 (1): 32–41.

Schalow, F. (2000), 'Who speaks for the animals?: Heidegger and the question of animal welfare', *Environmental Ethics*, 22 (3): 259–71.

Schmidt, A. (2014), 'Context-aware computing: Context-awareness, context-aware user interfaces, and implicit interaction', in M. D. Soegaard and R. F. Aarhus (eds), *The Encyclopedia of Human-Computer Interaction*, n.p., Denmark: The Interaction Design Foundation.

Schwanen, T. (2015), 'Beyond instrument: Smartphone app and sustainable mobility', *EJTIR*, 15 (4): 675–90.

Sensory Perspective (2012), 'MW1 THe Electrosmog Detector', *Detect Protect*. Available online: http://www.detect-protect.com/spcom-dwh/ccp51/cgi-bin/cp-app.cgi?usr=51F6230256&rnd=2100470&rrc=N&affl=&cip=&act=&aff=&pg=prod&ref=ELEKTROSMOG1&cat=WIRELESS-ENVIRONMENT&catstr=HOME:RF_CONSUMER:WIRELESS-ENVIRONMENT (accessed 20 December 2015).

Shadbolt, P. (2014), 'FireChat in Hong Kong: How an app tapped its way into the protests', *CNN*. Available online: http://edition.cnn.com/2014/10/16/tech/mobile/tomorrow-transformed-firechat/ (accessed 1 March 2017).

Shaviro, S. (2014), *The Universe of Things: On Speculative Realism*, Minneapolis: University of Minnesota Press.

Shaw, I. G. R. (2016), *Predator Empire: Drone Warfare and Full Spectrum Dominance*, Minneapolis: University of Minnesota Press.

Silverman, J. (2016), 'Just How "Smart" Do You Want Your Blender to Be?', *The New York Times Magazine*. Available online: https://www.nytimes.com/2016/06/19/magazine/just-how-smart-do-you-want-your-blender-to-be.html (accessed 1 March 2017).

Simondon, G. (1970), 'Entretien sur la mecanologie', *Youtube*. Available online: http://www.youtube.com/watch?v%C2%BCeXDtG74hCL4 (accessed 10 September 2015).

Simondon, G. (2009), 'Technical mentality', *Parrhesia*, 17: 17–27.

Simondon, G. (2017), *On the Mode of Existence of Technical Objects*, Minneapolis: Univocal Press.

Simpson, P. (2009), '"Falling on deaf ears": A postphenomenology of sonorous presence', *Environment and Planning A*, 41 (11): 2556–75.

Simpson, P. (2015), 'What remains of the intersubjective?: On the presencing of self and other', *Emotion, Space and Society*, 14: 65–73.

Smith, D. (2016), 'A Tesla driver was caught sleeping on the highway with his car on Autopilot', *Business Insider*. Available online: http://uk.businessinsider.com/tesla-autopilot-driver-caught-sleeping-2016-5?r=US&IR=T (accessed 2 March 2017).

Smith, M. R., and L. Marx (1994), *Does Technology Drive History?: The Dilemma of Technological Determinism*, Cambridge, MA: MIT Press.

Starosielski, N. (2015), *The Undersea Network*, Durham: Duke University Press.

Stiegler, B. (1993), 'Questioning technology and time', *Tekhnema*, 1: 31–44.

Stiegler, B. (1998), *Technics and Time, 1: The Fault of Epimetheus*, Stanford, CA: Stanford University Press.

Stiegler, B. (2003), 'Technics of decision an interview', *Angelaki*, 8 (2): 151–68.

Stiegler, B. (2007), 'Anamnesis and Hypomnesis', *Ars Industrialis*. Available online: http://www.arsindustrialis.org/anamnesis-and-hypomnesis (accessed 22 May 2010).

Stiegler, B. (2009), *Technics and Time, 2: Disorientation*, Stanford, CA: Stanford University Press.

Stiegler, B. (2010a), *For a New Critique of Political Economy*, Cambridge: Polity.

Stiegler, B. (2010b), 'The industrial exteriorisation of memory', in W. Mitchell and M. Hansen (eds), *Critical Terms for Media Studies*, 64–87, Chicago: University of Chicago Press.

Stiegler, B. (2010c), 'New industrial temporal objects', in R. Earnshaw and R. Guedj (eds), *Frontiers of Human-Centred Computing, Online Communities and Virtual Environments*, 450–60, London: Springer Verlag.

Stiegler, B. (2010d), *Technics and Time, 3: Cinematic Time and the Question of Malaise*, Stanford: Stanford University Press.

Stiegler, B. (2012), 'Die Aufklärung in the Age of Philosophical Engineering', *Computational Culture: A Journal of Software Studies*, 2: n.p.

Strengers, Y. (2013), *Smart energy technologies in everyday life: Smart Utopia?*, Berlin: Springer.

Sturken, M. (2008), 'Memory, consumerism and media: Reflections on the emergence of the field', *Memory Studies*, 1 (1): 73–8.

Sun. (2016), 'Sun – weather forecast powered by Dark Sky', *iTunes*. Available online: https://itunes.apple.com/us/app/sun-weather-forecast powered/id1161251105?ls=1&mt=8 (accessed 1 March 2017).

Sung, J.-Y., L. Guo, R. E. Grinter and H. I. Christensen (2007), '"My Roomba is Rambo": intimate home appliances', in J. Krumm, G. Abowd, A. Seneviratne, T. Strang (eds), *UbComp 2007: Ubiquitous Computing*, 145–63, Berlin: Springer.

Tamburrini, G. (2009), 'Robot ethics: A view from the philosophy of science', *Ethics and Robotics*, 11–22.

Tantsissa. (2017), 'HeartWatch', *Tantsissa*. Available online: http://heartwatch.tantsissa.com/home (accessed 1 March 2017).

Taub, E. (2016), 'Can Tesla's Autopilot Be Trusted? Not Always', *New York Times*. Available online: http://www.nytimes.com/2016/09/24/automobiles/autoreviews/can-teslas-autopilot-be-trusted-not-always.html (accessed 1 March 2017).

Tavani, H. T. (2011), *Ethics and technology: Controversies, questions, and strategies for ethical computing*, London: John Wiley & Sons.

Tesla. (2016a), 'Tesla Owners in Germany', *Tesla*. Available online: http://www.tesla.com/en_GB/blog/tesla-owners-germany (accessed 10 February 2017).

Tesla. (2016b), 'Upgrading autopilot: Seeing the world in radar', *Tesla*. Available online: http://www.tesla.com/en_GB/blog/upgrading-autopilot-seeing-world-radar (accessed 10 February 2017).

Tesla. (2017), 'Autopilot', *Tesla*. Available online: http://www.tesla.com/en_GB/autopilot (accessed 10 February 2017).

Thielmann, T. (2010), 'Locative media and mediated localities', *Aether: The Journal of Media Geography*, 5 (1): 1–17.

Thompson, M. (2017), *Beyond Unwanted Sound: Noise, Affect and Aesthetic Moralism*, New York: Bloomsbury Publishing.

Thomson, J. J. (1985), 'The trolley problem', *The Yale Law Journal*, 94 (6): 1395–1415.

Thrift, N. (2004), 'Remembering the technological unconscious by foregrounding knowledges of position', *Environment and Planning D: Society and Space*, 22 (1): 175–90.

Transport for London (2016), 'Private hire regulations review part two consultation report', *Transport for London*. Available online: https://consultations.tfl.gov.uk/tph/private-hire-proposals/ (accessed 2 May 2016).

Tsui, L. (2015), 'The coming colonization of Hong Kong cyberspace: government responses to the use of new technologies by the umbrella movement', *Chinese Journal of Communication*, 8 (4): 1–9.

Tuters, M., and K. Varnelis (2006), 'Beyond locative media: Giving shape to the internet of things', *Leonardo*, 39 (4): 357–63.

Uber. (2016), 'Uber.com homepage', *Uber*. Available online: http://www.uber.com/ (accessed 10 February 2017).

Urry, J. (2005), 'The complexity turn', *Theory, Culture & Society*, 22 (5): 1–14.

Van den Bulck, J. (2015), 'Sleep apps and the quantified self: Blessing or curse?', *Journal of Sleep Research*, 24 (2): 121–3.

Varela, F. J. (1999), 'The specious present: A neurophenomenology of time consciousness', *Naturalizing Phenomenology: Issues in Contemporary Phenomenology and Cognitive Science*, 64: 266–329.

Verbeek, P.-P. (2008), 'Morality in design: Design ethics and the morality of technological artifacts', in P. Vermaas, P. Kroes, A. Light and S. Moore (eds), *Philosophy and Design: From Engineering to Architecture*, 91–103, Berlin: Springer.

Verbeek, P.-P. (2009), 'Ambient intelligence and persuasive technology: The blurring boundaries between human and technology', *Nanoethics*, 3 (3): 231–42.

Virilio, P. (2007), *The Original Accident*, Cambridge: Polity.

Vitale, C. (2014), *Networkologies: A Philosophy of Networks for a Hyperconnected Age – a Manifesto*, Winchester, UK: Zero Books.

Washick, B., E. Wingrove, E. K. Ferguson and J. Bennett (2015), 'Politics that matter: Thinking about power and justice with the new materialists', *Contemporary Political Theory*, 14 (1): 63–89.

Waymo. (2017), 'Journey', *Waymo*. Available online: https://waymo.com/journey/ (accessed 2 February 2017).

Webster, M. (2014), 'Homeostasis', *Merriam Webster*. Available online: https://www.merriam-webster.com/dictionary/homeostasis (accessed 10 February 2016).

Wendling, A. (2009), *Karl Marx on Technology and Alienation*, London: Palgrave Macmillan.

West, K. E., M. R. Jablonski, B. Warfield, K. S. Cecil, M. James, M. A. Ayers, J. Maida, C. Bowen, D. H. Sliney and M. D. Rollag (2011), 'Blue light from light-emitting diodes elicits a dose-dependent suppression of melatonin in humans', *Journal of Applied Physiology*, 110 (3): 619–26.

Wetherell, M. (2012), *Affect and Emotion: a new social science understanding*, London: Sage Publishing.

Whitehead, A. N. (2010), *Process and Reality*, New York: Free Press.

Whitwam, R. (2014), 'How Google's self-driving cars detect and avoid obstacles', *Extreme Tech*. Available online: http://www.extremetech.com/extreme/189486-how-googles-self-driving-cars-detect-and-avoid-obstacles (accessed 5 February 2017).

Wilken, R. (2008), 'Mobilizing place: Mobile media, peripatetics, and the renegotiation of urban places', *Journal of Urban Technology*, 15 (3): 39–55.

Williams, N. (2016), 'Creative processes: From interventions in art to intervallic experiments through Bergson', *Environment and Planning A*, 48 (8): 1549–64.

Wilmott, C. (2016), 'Small moments in Spatial Big Data: Calculability, authority and interoperability in everyday mobile mapping', *Big Data & Society*, First published date: 19 September 2016, DOI: 10.1177/2053951716661364.

Wilmott, C. (2017), 'In-between mobile maps and media movement', *Television & New Media*, 18 (4): 320–35.

Wilson, M. W. (2014), 'Continuous connectivity, handheld computers, and mobile spatial knowledge', *Environment and Planning D: Society and Space*, 32 (3): 535–55.

Wilson, M. W., and M. Graham (2013), 'Situating neogeography', *Environment and Planning A*, 45 (1): 3–9.

Winner, L. (1980), 'Do artifacts have politics?', *Daedalus*, 109 (1): 121–36.

Wolfsfeld, G., E. Segev and T. Sheafer (2013), 'Social media and the Arab spring: Politics comes first', *The International Journal of Press/Politics*, 18 (2): 115–37.

Wood, A. W. (2006), 'How dangerous are mobile phones, transmission masts, and electricity pylons?', *Archives of Disease in Childhood*, 91 (4): 361–6.

Zook, M., and M. Graham (2007), 'The creative reconstruction of the internet: Google and the privatization of cyberspace and digiPlace', *Geoforum*, 38 (6): 1322–43.

INDEX